# The Global Energy Handbook

## *Understanding the Flow and Use of Global Energy*

The Basics:

*Where Energy Comes From*

*Where Energy Goes*

*Energy and Dollars*

*Alternative Sources*

Nothing Here On:

*Global Warming*

*When Do We Run Out*

*How to Save Energy*

John R. Fortun

The Global Energy Handbook

ISBN: 978-1-935125-10-5



To order additional books, go to:

www.RP-Author.com/Fortun

Printed in the United States of America

Robertson Publishing
59 N. Santa Cruz Avenue, Suite B
Los Gatos, California 95030 USA
(888) 354-5957 • www.RobertsonPublishing.com

# Preface

***This handbook will change the way you think about global energy.***

*The River of Energy*

First, you will begin to think of global energy as an immense, surging river with many input streams and an endless number of *final* destinations – *final* in that all energy, regardless of form, will forever be lost upon reaching its many and varied end points. You will see that the *River of Global Energy* is like no other river – it is always rising.

*No More Confusing Energy Terms*

Next, you will begin to think of energy units in apples-to-apples terms. When you read an article that contains mixed units – barrels of oil, megawatts, megajoules and therms – you will want to know what this means in terms of a common and easily understood set of energy units. This handbook provides a consistent and easy-to-use set of units that will help you interpret the often confusing and incomplete information on energy found in the media, in technical papers and on the Internet.

*You Will Question All Energy Information*

You will question all energy-related information that comes your way: "Am I reading about *prime source energy* or *end-use energy*?" or "How significant is that energy savings when looked at from a global perspective?"

You will begin to carry your copy of the *Global Energy Flow Chain* with you everywhere you go. You will pull it out at work or at cocktail parties whenever someone makes an uninformed statement on energy.

## Preface

*No More Band-Aid Solutions*

You will no longer look at energy as something with a fixed set of problems that can be treated with simple band-aid solutions. Instead, you will think of energy as a dynamic and changing force that defies our attempts to limit or control it. You will see that energy does not hold still while we try to fix it. You will understand why energy-saving solutions that appear so promising today won't hold up under the relentlessly increasing energy demands expected over the next several decades.

*Thinking Energy at the Global Level*

Finally, you will begin thinking of energy issues on a truly global level. You will still be concerned about the energy you spend at home and on your car. You will still try to conserve as much as possible. You will still be concerned about the energy we consume as a nation and our dependence on foreign oil, but you will think of these issues from a far more global viewpoint than you ever did before. You will understand what I mean when I say we have been looking at our energy problems through the wrong end of the telescope.

# Table of Contents


| | |
|---|---|
| **Preface** | **ii** |
| **Tables of Contents** | **iv** |
| **Figures and Tables** | **viii** |
| **Abstract** | **xii** |
| | |
| **Introduction** | **I-1** |
| | |
| **Section 1: What's This Handbook All About?** | **1-1** |
| A Quick Reference | 1-2 |
| What's Not Here | 1-2 |
| Confusion and Pitfalls | 1-3 |
| End-Use Consumption | 1-6 |
| Percentage of What? | 1-7 |
| Land of the Lost | 1-7 |
| Data-Collection Approach | 1-8 |
| Internet Sources | 1-8 |
| The Reference Year | 1-9 |
| The Global Energy Flow Chain – A Preview | 1-10 |
| Summary of Section 1 | 1-11 |
| | |
| **Section 2: The Sources of Global Energy** | **2-1** |
| Energy Information – A Mixed Bag | 2-1 |
| The Energy Lineup | 2-2 |
| The Alternatives | 2-3 |
| The Energy Heavyweights | 2-4 |
| It's All About BTU | 2-4 |
| Global Energy in BTU | 2-6 |
| Looking Back to 1980 | 2-9 |
| Alternative Energy Growth | 2-10 |
| Global Energy Growth | 2-11 |
| Summary of Section 2 | 2-12 |

# Table of Contents

**Section 3: Where Does All the Energy Go?** **3-1**
The Energy Flow Chain 3-1
Energy Flow to the Sectors 3-5
End-Use Consumption by the Sectors 3-11
The Energy Flow Chain – The Big Picture 3-16
How Much Goes Into? 3-17
End-Use Consumption Growth Rate 3-18
Prime Source Energy by the Sectors 3-20
Summary of Section 3 3-21

**Section 4: Energy in Everything** **4-1**
Energy Content of the Prime Sources 4-1
Energy Reference Tables 4-2
Petroleum: Mr. Big (for now) 4-3
Coal: The Energy Workhorse 4-8
Natural Gas: Low-Profile Worker 4-10
Nuclear Energy: The Silent Worker 4-11
Hydroelectric Energy: Water Over the Dam 4-12
Biofuels: Home Grown 4-13
Wind Energy: The Big Breeze 4-14
Geothermal Energy: Deep Heat 4-14
Solar Energy: Ray of Hope? 4-15

**Section 5: Energy on the Home Front** **5-1**
The Cost of Energy 5-1
The Electricity Flow Chain 5-3
Energy for Electricity Generation 5-4
Electricity End-Use Consumption 5-7
Electricity Consumption by the Sectors 5-8
What Does All This Really Cost? 5-10
Residential Energy Consumption by Item 5-13
Household Energy Use – The Global Picture 5-15
Summary of Section 5 5-17

## Table of Contents

**Section 6: Energy and Your Automobile** **6-1**

Vehicle Fuel Efficiency – Miles per BTU 6-1

Calculating Vehicle Efficiency – Two Approaches 6-3

Auto Vehicle Efficiency Chains 6-11

Energy Costs by the Mile 6-14

Thoughts on Mileage 6-18

Thoughts on Improved Mileage 6-19

Summary of Section 6 6-24

**Section 7: Looking Back at Energy Savings** **7-1**

Energy Wake-Up Calls 7-1

How Much Has All of This Helped? 7-3

What, Me Waste Energy? 7-5

What Is Energy Waste? 7-6

Summary of Section 7 7-6

**Section 8: The Alternative Energy Sources** **8-1**

Near-Term vs. Long-Term Energy Solutions 8-1

Solar Energy Generation 8-2

Wind Energy Generation 8-5

Biofuel Energy Production 8-6

Hydrogen and the Fuel Cell 8-10

Geothermal Energy Generation 8-11

Summary of Section 8 8-11

## Table of Contents

**Section 9: How to Use This Handbook** **9-1**

Some "What-If" Energy-Saving Examples 9-1

Example 1: Fluorescent Bulbs 9-1

Example 2: Clothes Dryers 9-2

Example 3: Residential Electricity 9-3

Example 4: Reduced Driving 9-4

Example 5: Hybrid Vehicles 9-5

Example 6: Electric Vehicles 9-9

Example 7: Proper Tire Inflation 9-12

Summary of Section 9 9-13

**Section 10: The Energy Triad** **10-1**

Global Energy Consumption – Past and Future 10-1

Energy and "The Economy" 10-2

Fossil-Fuel Growth 10-3

Greenhouse Emissions 10-4

The Gross World Product, GWP 10-5

The Energy Triad Plotted 10-6

What's Not the Solution? 10-7

Another Game to Play 10-9

Closing Statement 10-10

**Section 11: Rerference Data and Figures** **11-1**

General References 11-1

Reference Figures and Data Reductions 11-3

Data-Reduction Examples 11-3

# Figures and Tables


| | | |
|---|---|---|
| **Section 1: What's This Handbook All About?** | | **1-1** |
| Figure 1-1 | Confused About Energy? | 1-1 |
| Table 1-1 | Nothing Included Here On: | 1-2 |
| Figure 1-2 | More Confusing Issues | 1-3 |
| Table 1-2 | Summary of Ground Rules and Definitions | 1-10 |
| **Section 2: The Sources of Global Energy** | | **2-1** |
| Figure 2-1 | World Prime Energy Sources by Percent | 2-2 |
| Table 2-1 | World Prime Energy Sources Production | 2-7 |
| Figure 2-2 | World Prime Energy Source Energy by QBTU | 2-8 |
| Figure 2-3 | World Prime Energy Source Energy Fuel Type | 2-9 |
| Figure 2-4 | Alternative Energy by Percent PSE | 2-10 |
| Figure 2-5 | Alternative Energy Growth in QBTU | 2-10 |
| Figure 2-6 | World Prime Source Energy Growth | 2-11 |
| Figure 2-7 | EIA World Energy Growth Scenarios | 2-12 |
| **Section 3: Where Does All the Energy Go?** | | **3-1** |
| Figure 3-1 | The Energy Flow Chain | 3-3 |
| Figure 3-2 | The Energy Flow Chain in QBTU | 3-4 |
| Figure 3-3 | Energy Flow Chain to the Sectors | 3-5 |
| Table 3-1 | The End-Use Energy Consumption Sectors | 3-7 |
| Figure 3-4 | End-Use Consumption by the Sectors | 3-11 |
| Table 3-2 | World End-Use Consumption by Sector | 3-12 |
| Figure 3-5 | World End-Use Consumption (%) by Sub-Sectors | 3-13 |
| Figure 3-6 | World End-Use Consumption (QBTU) Sub-Sectors | 3-14 |
| Figure 3-7 | The Global Energy Flow Chain | 3-15 |
| Figure 3-8 | World End-Use Consumption by Sector | 3-19 |
| Table 3-3 | World End-Use Consumption by Year | 3-19 |
| Figure 3-9 | Prime Source Energy Use by the Sectors | 3-19 |

## Figures and Tables


**Section 4: Energy in Everything** 4-1

Figure 4-1 Energy Content Comparison 4-1

Table 4-1 Petroleum 4-3

Figure 4-1a Where a Barrel of Oil Goes 4-4

Figure 4-1b World Petroleum Production 4-5

Figure 4-1c World Oil Consumption by Sector 4-5

Table 4-1a Gasoline 4-6

Figure 4-1d World Oil and Gasoline Production 4-6

Table 4-1d World Oil Production 4-7

Table 4-1e World Gasoline Production 4-7

Table 4-2 Coal 4-8

Figure 4-2a World Coal Production 4-8

Figure 4-2b Variations in World Coal Data 4-9

Table 4-3 Natural Gas 4-10

Table 4-3a Natural Gas Conversions 4-10

Figure 4-3 World Natural Gas Production 4-11

Table 4-4 Nuclear Energy 4-11

Figure 4-4 World Nuclear Energy Growth 4-12

Table 4-5 Hydroelectric Energy 4-12

Figure 4-5 World Hydroelectric Energy 4-13

Table 4-6a Ethanol 4-13

Table 4-6b Biomass 4-14

Table 4-7 Wind Energy 4-14

Table 4-8 Geothermal Energy 4-14

Table 4-9 Solar Energy 4-15

## Figures and Tables


**Section 5: Energy on the Home Front** **5-1**

| | | |
|---|---|---|
| Figure 5-1 | Energy Costs per Million BTU | 5-1 |
| Figure 5-1a | Consumer Energy Costs | 5-1 |
| Table 5-1 | Residential Energy-Users | 5-2 |
| Figure 5-2 | World Energy Flow Chain | 5-3 |
| Figure 5-3 | World Energy (PSE) to Electricity Generation | 5-4 |
| Figure 5-4 | PSE Input to Electricity Generation | 5-7 |
| Figure 5-5 | World Electricity Consumption by EUC | 5-8 |
| Figure 5-6 | World Electricity Consumption by Sectors | 5-9 |
| Table 5-2 | Electricity Units and Abbreviations | 5-9 |
| Figure 5-7 | Electricity Rates Adjusted for Inflation | 5-11 |
| Figure 5-8 | Natural Gas Rates Adjusted for Inflation | 5-12 |
| Figure 5-9 | Residential EUC Estimation | 5-13 |
| Figure 5-10 | Your Real Energy Consumption | 5-14 |
| Figure 5-11 | Energy Use at the Global Level | 5-16 |
| Figure 5-12 | Residential Energy Flow Chain | 5-16 |

**Section 6: Energy and Your Automobile** **6-1**

| | | |
|---|---|---|
| Figure 6-1 | Vehicle Miles per Million BTU by EUC | 6-4 |
| Figure 6-2 | Vehicle Miles per Million BTU by PSE | 6-5 |
| Table 6-1 | Fuel Vehicle Parameters "Tank to Wheel" EUC | 6-6 |
| Table 6-2 | EV Parameters "Battery to Wheel" EUC | 6-7 |
| Figure 6-3a | EV Battery Specific Energy | 6-8 |
| Figure 6-3b | EV Battery Specific Energy Compared to Gasoline | 6-9 |
| Table 6-3 | $H_2$ Vehicle "Tank to Wheel" EUC | 6-10 |
| Figure 6-4 | Auto Vehicle Efficiency Chains | 6-13 |
| Figure 6-5 | Cost Per Mile by Vehicle Type | 6-14 |
| Figure 6-6 | Energy to Run Your Car vs. Your Body | 6-15 |
| Figure 6-7 | Gasoline Cost Per Gallon Adjusted for Inflation | 6-16 |
| Figure 6-8 | Gasoline Cost Adjusted for Inflation and Mileage | 6-17 |

## Figures and Tables


| | | |
|---|---|---|
| Figure 6-9 | Motor Vehicle Miles per Gallon History | 6-19 |
| Figure 6-10 | Average Miles Driven per Year | 6-21 |
| Figure 6-11 | Gasoline Consumption History | 6-21 |
| Figure 6-12 | World Motor Vehicle Growth | 6-22 |
| Figure 6-13 | The Petroleum Energy Flow Chain | 6-23 |

**Section 7: Looking Back at Energy Savings** **7-1**

| | | |
|---|---|---|
| Figure 7-1 | World Prime Source Energy Growth | 7-1 |
| Table 7-1 | Energy Savings Initiatives Since 1970 | 7-2 |

**Section 8: The Alternative Energy Sources** **8-1**

(No tables or figures)

**Section 9: How to Use This Handbook** **9-1**

| | | |
|---|---|---|
| Figure 9-1 | Petroleum Energy Flow Chain – Hybrids | 9-6 |
| Figure 9-2 | Energy Savings from Improved Mileage | 9-8 |
| Figure 9-3 | Petroleum Energy Flow Chain – 50% EVs | 9-14 |

**Section 10: The Energy Triad** **10-1**

| | | |
|---|---|---|
| Figure 10-1 | Energy Consumption and Savings Initiatives | 10-1 |
| Figure 10-2 | The Energy Triad – A Precarious Balance | 10-2 |
| Figure 10-3 | Global Fossil-Fuel Energy Production | 10-3 |
| Figure 10-4 | Global $CO_2$ Emissions | 10-4 |
| Figure 10-5 | Normalized Fossil-Fuel and $CO_2$ Emissions | 10-5 |
| Figure 10-6 | Gross World Product, GWP | 10-5 |
| Figure 10-7 | Normalized Global Energy, $CO_2$ Emissions and GWP | 10-6 |

**Section 11: Reference Data and Figures** **11-1**

(Selected EIA Reference Figures)

Abstract

The *Global Energy Handbook* describes world energy from initial production to end-use consumption. The handbook is unique in that it does this using a clear and consistent set of energy units, the BTU. For many people energy information might as well be written in a foreign language. In most articles energy terms such as megawatts, barrels of oil and gallons of gasoline are so intermixed the reader cannot see what's happening at the global level.

The handbook, compiled mainly from Internet and current media sources, is organized and presented in a way that can easily be followed by those with or without technical backgrounds. The handbook is unique in that it does not address the usual subjects of energy technology, global warming or "ways to save energy" – the literature being rife with this sort of information.

Upon gaining a feeling for the global energy flow process, the reader will begin to question the conventional wisdom on energy conservation, efficiency and the alternative energy sources that routinely appears in the media. Numerous *What If* examples are given to illustrate why the conventional wisdom has been unable to reduce or even slow down global energy consumption. Two of the examples given in the handbook are:

*What if everyone in the world drove a fuel-efficient automobile?*

*What if every household in the world switched to fluorescent bulbs?*

You will be surprised and dismayed by what these examples tell us about the difficulty of saving energy at the global level. They will make clear the meaning of two key phrases used throughout the handbook:

*"Energy does not hold still while we try to fix it."* and

*"We have been looking at energy through the wrong end of the telescope."*

This handbook should be of great value to students of the energy and environmental sciences, to those in the media reporting on energy issues and to average citizens who just want to better understand the flood of energy information coming their way.

# Introduction

## A Little Background

The purpose of this handbook is to provide the reader with a basic, but essential understanding of the overall global energy-flow process. With this understanding he or she will be able to ask critical questions on current energy issues and technologies. More importantly, the reader will be able to challenge the often misleading claims on energy-saving proposals encountered daily in the media or in technical literature.

### Why I Wrote This Handbook

Every time I read an article or watched a TV program on energy, it would usually leave me with more questions than answers. I became tired of using the Internet or going back to my old engineering texts to answer the reoccurring question of "What does this mean from the global level?" I decided to compile a set of basic energy data and conversion factors to keep at my fingertips so that I could quickly address questions like this.

### What This Handbook Does

This handbook tracks the flow of global energy from its initial production to its end-use consumption. This is done by expressing all forms of energy in a consistent set of understandable units. Those seeking to better understand global energy are confronted with a myriad of mixed and confusing energy units making it very difficult to do any meaningful comparisons, let alone arrive at an overall understanding of the global energy-flow process. The energy-flow process, from initial production to final use, is described in straightforward text along with easily understood tables and diagrams.

### Who Will Find This Handbook Useful?

This book should be invaluable to students in the areas of environmental and energy sciences, to those in the media reporting on energy issues and to average citizens who just want to better understand the newspaper, magazine and television coverage of the energy-related issues they routinely hear about.

# Introduction

## Prerequisites

Do I need a technical background to understand this handbook? No, you will only need to have a real concern for energy issues plus a desire to go beyond the conventional wisdom for solving our energy problems that is usually given in the media.

## What's Not Covered

Almost everything published today on energy falls into the categories of global warming, conservation and energy savings or alternative energy technology. None of these are covered in this handbook. The Internet, the media and technical literature have more than covered the bases on these subjects.

## More Than Just a Data Handbook

The original intent of this handbook was to compile a consistent and easy-to-use set of data on global energy. Energy information, for most people, might as well be written in a foreign language. Energy units like megawatts, barrels of oil and gallons of gasoline often appear in the same article with confusing profusion. A statement like "Solar energy capacity increases by 1000 megawatts" means nothing to most of us, unless it can be related to some known energy-reference point; statements such as this continually appear in the media without interpretation.

As the handbook took shape, I began to realize that most of the conventional wisdom we hear about global energy – we can solve our energy problems simply through conservation, by being more energy efficient and by investing more in alternative energy technologies – did not ring true when viewed from past history and the global perspective.

# Introduction

This has been the mantra over the last four decades since the *Energy Crises* of the 1970s. Despite being endlessly repeated in newspaper editorials and Sunday supplement articles, global energy consumption has continued on its relentless rise.

You will not become an expert on energy after reading this book, but you will gain an understanding of the global energy-flow process that many in the media appear not to have.

## The Wrong End of the Telescope

It will become clear why conserving energy at home, driving a more fuel-efficient car, buying *green* products and in general being more energy conscious – all things I fully recommend and practice – will not solve our global energy problem.

*We have been looking at the problem through the wrong end of the telescope.* We need to turn the telescope around to see why the conventional wisdom has not been working. This handbook, while offering no solutions, should help you in understanding the magnitude and gravity of our global energy problem.

# Introduction

## By the Sections

### What Is This Handbook All About? – Section 1

The ground rules and assumptions used throughout this handbook are given in Section 1. Many of the pitfalls and points of confusion encountered when trying to understand global energy information are pointed out in this section. This section also gives an introduction to the *Global Energy Flow Chain*, a tool you should find very useful for interpreting energy information.

### The Sources of Global Energy – Section 2

The prime sources of global energy are described in Section 2 using a consistent set of energy units – the *British Thermal Unit* or BTU. You will see why the fossil fuels dominate global energy use. You will see how the alternative energy sources (biofuel, solar, wind, etc.) are fairing against the fossil fuels. This may surprise you from everything you have been hearing. Understanding P*rime Source Energy* or PSE is the main focus of this section.

### Where Does All the Energy Go? – Section 3

Here you will see where global prime source energy is used in terms of the major end-use consumption sectors. You will see that a relatively small amount of energy is consumed by the residential sector (your utility bills) compared to the other energy consumption sectors. You will be surprised at the small percentage of electricity that reaches you after its journey from the generation plant to your home receptacle. In this section you will become very familiar with the *Global Energy Flow Chain*, the most important and useful figure in this handbook. The key term in Section 3 is *End-Use Consumption* or EUC.

# Introduction

### Energy in Everything – Section 4

Pertinent data for each of the prime energy sources is given in Section 4 in tabular form. This may sound a little boring, but these tables can be very useful for making quick and easy comparisons between the several prime source energy types – something you may want to do after gaining an appreciation of the overall energy-flow process.

### Energy on the Home Front – Section 5

Energy consumption at the household level is the subject of Section 5. Here it is seen that household energy is only 15 percent of all consumed energy. Despite this, an endless number of articles scolding you for wasting energy in and around your home routinely appear in newspapers and magazines. You will see that the average family uses several times more energy on consumer goods and related services than directly on household utilities.

### Energy and Your Automobile – Section 6

Energy consumption and your car is the topic of Section 6. Here energy-efficiency comparisons are made between several motor vehicle types. You will see how vehicle efficiencies differ when comparisons are made between *prime-source energy* and *end-use energy*. The cost of gasoline is looked at from an inflation adjusted viewpoint – this may surprise you, but it won't make you feel any better at the gas pump. Finally, you will see why the improvements in automobile fuel-efficiency over the past thirty years have failed to reduce global gasoline consumption.

# Introduction

## Looking Back at Energy Savings – Section 7

In Section 7 it will be seen why the many energy savings initiatives introduced and promoted over the last several decades have not been able to slow the consumption of global energy. This is explained in terms of the *Global Energy Savings Shrink Factor* – savings which appear so promising at the lower levels of the energy consumption chain shrink in percentage when translated to the global level.

## Near-Term and Far-Term Solutions

Before summarizing the alternative energy sources, the use of *near term* and *far term* needs to be clarified. Near-term energy solutions, as used here, apply to the next 25 years or so when the double whammy of diminished petroleum supplies combine with the possibility of irreversible global-warming effects (If you don't believe in global warming, don't stop reading yet. The information given here is fundamental, and in itself has nothing to do with opinions on global warming).

Far-term energy solutions are probable for our grandchildren or children's grandchildren, but are unlikely to help us with the upcoming energy crunch. The question the reader will learn to ask in any discussion on energy solutions is: "Is this a near-term or a far-term solution?" Far-term solutions will also be referred to in this handbook as *"time-tunnel-escape-hatch"* solutions or just as time-tunnel solutions.

In most discussions (arguments) on global energy, the distinction between near and far-term solutions is usually the first thing to go. What is the issue? Is it fossil-fuel consumption and global warming (near-term) or is it the issue of providing energy for future generations (far-term)?

# Introduction

**The Alternative Energy Sources – Section 8**

In Section 8 the *Alternative Energy s*ources will be looked at in terms of their inherent capabilities to provide global energy, not their technological limitations which have been the subject of countless articles and research papers. You will see why the alternative sources can never be more than a partial solution to our energy problems – this contrary to all of the media and political rhetoric.

Biofuels (ethanol and biodiesel), as you have been reading, are already causing food supply and environmental concerns even though they currently provide less than 0.25 percent of global energy. Cellulosic biofuels, with the promise of being grown on semi-arid land at very low cost, may get past some of these problems, but for now they remain in the time-tunnel category.

Solar and wind energy, both receiving great attention as examples of alternatives needing further investment, are limited by two fundamental factors. First, they only produce electricity – less than 40 percent of all energy use. Secondly, they are part-time workers requiring the continual back up from conventional (mostly fossil fuel) energy sources.

Yes, we most certainly should increase our investments in the alternative energy sources, but with a realistic understanding of their inherent limitations.

**How to Use This Handbook – Section 9**

In Section 9 several "What If" energy-saving proposals are looked at in terms of savings at the global level. What if every household in the world switched to fluorescent bulbs? What if everyone in the world reduced their driving by ten percent? What if everyone in the world switched to a high-mileage automobile? You will be surprised (dismayed) to see how these energy-saving ideas shrink in effectiveness when viewed from the global perspective.

## Introduction

You will see how conservation efforts, as well meaning as they may be, are quickly washed away by the ever increasing demand for global energy. Here are some examples of energy saving proposals when viewed from the global perspective.

*Saving Energy at Home*

Household energy – electricity and heating fuel – accounts for 15 percent of all consumed energy. If *every* household in the world were to reduce its energy use by 10 percent, global energy consumption would be reduced by 1.5 percent (Section 5).

*Fluorescent Light Bulbs*

Residential lighting accounts for about10 percent of household electricity, but at the global level this is only about two percent of all prime source energy. If *every* household in the world switched to fluorescent bulbs, global energy production would be reduced by about one percent (Section 9, Example 1).

*Reduced Driving*

Passenger vehicles consume nearly 500 billion gallons of gasoline yearly. If *everyone* in the world reduced their personal driving by 10 percent, global petroleum consumption would be reduced by 3.2 percent; global fossil-fuel consumption would be reduced by 1.4 percent (Section 6, Figure 6-13).

*Fuel-Efficient Vehicles*

Increasing automobile mileage from 22 mpg to 44 mpg should result in cutting gasoline consumption by 50 percent – at least according to conventional wisdom. However, this would be true only if the number of vehicles in the world remained constant during the 20 years or so it would take to turn over the global fleet to the higher mileage.

## Introduction

Automobile mileage increased from about 12 mpg in 1970 to about 22 mpg today, but gasoline consumption actually doubled during that time because the number of vehicles in the world quadrupled (200 to over 800 million). There is no indication of a slowdown in the world's motor vehicle population. By 2050 it is estimated that there could be two billion or more motor vehicles worldwide. If this proves to be true, gasoline consumption, along with oil consumption, will continue to rise despite our best attempts at improving fuel economy (Section 6).

### The Energy Triad – Section 10

A simplified look at the relationship between consumer spending, energy consumption and greenhouse gas emissions – *The Energy TRIAD* – is presented in Section 10. No surprises here, you will see that the elements of the TRIAD are more than closely related, they are essentially in lockstep.

Proposed solutions to the world's energy problems are not given in this final section, only a reminder of the purpose of the handbook – to provide the reader with a basic understanding of the global energy-flow process. With this, the reader should be able to interpret most of the energy information presented in the media. More importantly, the reader can critically question energy issues and challenge those proposed solutions, which appear so promising at first look, but just won't hold up under the ever increasing demand for energy.

### Reference Data and Figures – Section 11

This section gives the primary reference sources used in the handbook. Also several key figures and tables from the Energy Information Administration (EIA) are given in their original form. The author's graphical reductions of these EIA figures, as used in the body of the handbook, are given for comparative purposes.

## Section 1

## What Is This Handbook All About?

The purpose of this handbook is to provide you, the user of global energy, with the means to understand the often-confusing array of facts, figures and claims encountered when reading, watching TV specials or downloading energy information from the Internet. You know what I mean: barrels of oil, megawatts of power, megajoules of energy, kilowatt-hours of electrical energy or ergs of who knows what. Then there is the "how much energy could be saved if only everyone did such and such." This guide gives you the information you will need to put all of these varied energy inputs into a global perspective.

Those with technical backgrounds may accuse me of watering down this subject. Maybe so; however, I have forty years of experience in aerospace engineering dealing with the design, analysis and testing of complex systems where force, energy, power and efficiency are everyday considerations. When I see something on TV or read something in the Sunday newspaper on the subject of energy, I often find myself having to dig out my old mechanical engineering textbooks and energy conversion tables to understand its significance.

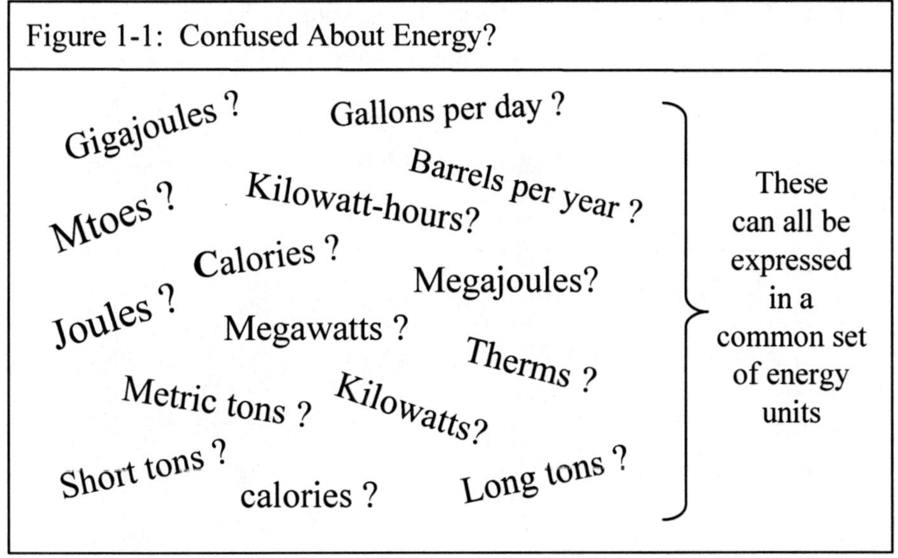


Figure 1-1: Confused About Energy?

## Section 1: What Is This Handbook All About?

### A Quick Reference

This handbook is set up primarily to be a quick reference for assessing energy data and related working parameters. There are countless references available describing in detail energy technology and how energy is produced. This handbook is intended to supplement these more comprehensive tutorial-type works. All energy-related parameters are expressed in a common set of energy units so that quick cross-checks and comparisons can be made. The *British Thermal Unit* (BTU) is used as the common-denominator energy unit. Whether we are talking about gasoline, natural gas, coal, solar cells, wind machines or any other form of energy, the BTU will be used to relate these sources of energy to each other.

### What's Not Here

You will not find anything on the threats of global warming or the impending energy crisis. Nor is there any preaching on the virtues of energy conservation. No lists of "100 Ways to Save Energy" are included. No theoretical formulations, energy equations or statistical data and analyses are needed to understand what's going on here. As you will see, everything is based on readily available energy information, simple arithmetic and cognitive knowledge. I have tried not to interject my personal opinions into this work. The production and use of the earth's energy resources will speak for themselves. The potential for the alternative energy sources and how they compare against fossil resources will also speak for themselves. The merits of the many energy-saving ideas that you routinely hear about will become self-evident.

| Table 1-1: Nothing Included Here On: |
|---|
| Global Warming – You shouldn't have to be warned by me on this. |
| Tips for Saving Energy – A standard in Sunday supplements for years. |
| Energy Technology – From the basics to Ph.D. papers abound on the Internet. |
| No Scolding You for Wasting Energy – We are all in this boat together. |

## Section 1: What Is This Handbook All About?

### Confusion and Pitfalls

Most, if not all, of the energy information and data provided here can easily be found on the Internet. But even if you are fairly familiar with energy units, energy terminology and conversion factors, the interpreting of what you have found could be a frustrating job. Here are some pitfalls and areas of possible confusion.

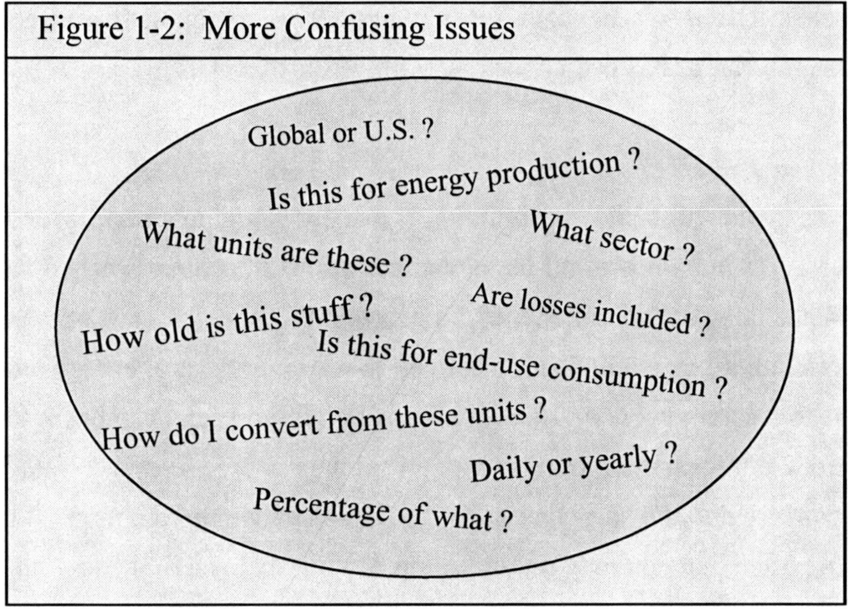


*Inconsistent or Undefined Units*

Inconsistent or undefined energy units are quite often encountered in tables or graphs. Are you looking at megawatts or kilowatt-hours or what? Sometimes you can figure this out by reading through the accompanying text, sometimes not. Often you will need to convert energy units in one section of a document to some other set of units so that you can make meaningful comparisons.

Occasionally after converting units, you may find disagreements between energy values given in different sections of the same report. At times authors simply forget to explicitly define their units. Other times data has been extracted from a larger work, and a clear definition of units has been left behind.

## Section 1: What Is This Handbook All About?

*Undated or Obsolete Data*

Undated or obsolete energy-related documents and data are common on the Internet. What at first appears to be a good set of information may be as much as ten years old. On many occasions information has been extracted from larger reports, but details, like the date, are nowhere to be found. Some reports will contain data from mixed years (oil from 2004, coal from 2002, hydroelectric from whenever). This may require the extrapolating of the data to a common year, as has been done many times for this work.

*Per Day or Per Year?*

Energy production and consumption is usually given on a per-day or per-year basis (as in barrels of oil per day or short tons of coal per year). In many documents these are so mixed that it's nearly impossible to make sense of what you are reading. You will need to convert to a common set of energy units as well as to a common energy-per-time basis before you can even begin to interpret the information. In this work all primary source reference data has been converted to the "per-year basis." You will encounter the term "BTU per year" so often that you may tire of seeing it. Still, that is a lot better than having to continually convert things like "gallons per day" or "therms per year" into "BTU per year" in order to understand what you are reading.

*Global or U.S. Information?*

A large portion of the energy information that appears on the Internet is for U.S. production, use and economic concern only. Quite often, though, it is not clear whether you are looking at something on the global level or something pertaining to the U.S. only. Again, by reading through the accompanying text you may be able to determine if it is U.S. or global information. If there is no text you may be out of luck unless you can relate the data to some known U.S. or global energy quantity that you can then use as a reference point.

## Section 1: What Is This Handbook All About?

*Alternative or Renewable?*

Often energy information will be grouped into the categories of fossil, renewable or alternative energy sources. In this work the prime energy sources will be discussed in terms of fossil and non-fossil fuels. The non-fossils include hydroelectric, nuclear and the family of alternative energy sources – solar, wind, geothermal, ethanol, biodiesel and biomass. The alternatives can be thought of as new (or rediscovered, as in the case of wind energy) technologies which are under development in the hopes of reducing global dependency on fossil fuels. The fossil fuels, of course, consist of petroleum, coal and natural gas. The term renewable will not be used as such, although it applies to the hydroelectric energy sources as well as to all of the alternative energy sources.

*Prime Source Energy*

Terms for describing the initial production of global energy versus the final or end-use phase can be confusing. The terms "energy production" or "energy generation" or "energy consumption" are often used interchangeably to describe the initial phase of energy production. In this work the term *Prime Source Energy* or PSE is used to describe the initial phase of global energy flow. Globally this means all energy production, regardless of where the raw resources may have originated. For individual countries, like the United States, prime source energy includes all energy produced internally plus all imported energy less any exported energy (yes, we do export some).

Here are some of the terms commonly used by other authors and references for the initial phase of global energy production:

"World Primary Energy Consumption"

"Marketed Energy Consumption"

"Marketed Energy Use"

"World Production of Prime Energy"

"World Energy Production"

## Section 1: What Is This Handbook All About?

When confronted with a new set of energy information, you first must determine if it pertains to the prime source stage or to the end stage of the energy-flow process. There will be much more about this in Section 2 (Sources of Global Energy). In the following sections you will become familiar with the magnitudes (in BTU) of all the prime energy sources. Of more importance, you will become familiar with global energy flow and the ever-increasing demands that keep global energy production and consumption on the continual rise.

### End-Use Consumption

The term *End-Use Consumption* or EUC is used in this work to describe the final phase in the energy-flow process. This is the end of the line for global energy. This is where industry, agriculture, transportation, business, government and you as an energy consumer spend all of the globally produced prime source energy. The subject of end-use consumption will be covered in Section 3 (Where Does All the Energy Go?).

The terms "energy consumption" or "energy delivered" or "consumer end-use" usually designate the end of the energy-flow process. I have encountered data labeled "Energy Consumption," which was actually data for prime source production. I have also run across data labeled "Energy Generation Consumption," which required a little backtracking to determine what was really meant.

The term "energy demand" is also used in multiple ways. Often demand is used when referring to energy production, and at other times when referring to end-use consumption. To avoid confusion we will try not to use the term "demand" unless in an explicitly defined situation, such as "the demand for global energy."

Here are some of the possibly confusing terms used to describe the end phase of the energy chain:

"World Delivered Energy"

"World Consumed Energy"

"World Energy Consumption"

"World Energy Demand"

## Section 1: What Is This Handbook All About?

**Percentage of What?**

Graphs and pie charts are frequently given which show percentages of energy use, but they are not always clear as to the percentage of what. If the percentage data you are looking at is simply labeled "Percentage of Energy Consumption," you will have to determine the following:

1. Is the data for percentages within a particular consumption sector?
2. Is it a percentage of total energy consumption (EUC)?
3. Is the data for U.S. or global consumption?
4. Is the data a percentage of total world energy production (PSE)?

As you will see in Section 3, there is a big difference between world energy production and world end-use consumption.

**Land of the Lost**

One of the more confusing things you will encounter when accessing energy information has to do with electricity consumption and energy losses. Energy losses, as you will see in later sections, occur throughout the electricity chain from initial generation to consumer end-use. Tables and graphs that give data on electricity consumption for the various end-use categories may not indicate if generation and transmission losses are included. It then gets more confusing when electricity consumption is mixed with direct-use energy (non-electricity as in oil, coal and natural gas) to give total energy consumption.

Only by having a good feeling for the overall energy flow from prime-source production to consumer end-use will you be able to interpret mixed energy information like this. In general, the labels on graphs or figures you encounter will be sufficiently ambiguous to defy your prompt interpretation. As you move through the sections of this work, you should attain the ability to determine where a set of data lies within the global energy chain, thus allowing you to make quick and easy interpretations. Here are some terms commonly used for the end phase of electricity consumption.

"World Net Electricity Consumption"

"World Net Generation of Electricity"

## Section 1: What Is This Handbook All About?

### Data-Collection Approach

The approach for this work was to compile energy information mainly from Internet sources, but supplemented with numerous articles and research studies. One can only arrive at a consensus set of values for production, consumption and all the other numerous world energy-related parameters. World energy data is collected by many different government agencies as well as by the major energy suppliers like British Petroleum. Each has different collection approaches as well as different definitions of what they are collecting. Data tables from the U.S. Energy Information Agency (EIA) often contain extensive footnotes telling what is included and what's not included, as well as detailed definitions of the terms used.

The major objective of this handbook is to provide an understanding of the flow of global energy from initial production (PSE) to end-use consumption (EUC). If you require the most recent and accurate data on a particular energy source, you will need to track this down elsewhere. The actual energy values and percentages given throughout this work are approximate and are of secondary importance to the understanding of the global energy-flow chain. As such then, this handbook should prove to be a valuable reference for the next several years even though the specific energy values given will undoubtedly change as the demand for global energy continues to grow.

### Internet Sources

The main Internet reference sources came from the U.S. Government's EIA Website (eia.doe.gov, etc). When searching for any energy-related information something from the EIA is sure to appear. There are so many EIA sites that it is difficult to keep track of which one you are looking at. The data may also be several years old, so be wary. Most EIA data is for the U.S., but EIA International data is also available, but this may not always be identified as such, especially in second-level tables or figures.

## Section 1: What Is This Handbook All About?

None of the figures, tables, diagrams used in the body of this document were taken directly from the reference sources. Instead, all information presented has been graphically reduced and is composite in nature and thus may not exactly agree with any particular reference source. Several original EIA reference figures and tables, along with the author's graphical reductions used in this work, are documented in Section 11 (Reference Figures and Data Reductions).

### The Reference Year

As much as possible, all energy data and related information has been normalized to the 2007-2008 time frame. Attempting to be more specific on the date proved to be an exercise in futility. Information dated "2007" could just as well be data for the year 2006. For example, data from a particular source labeled as 2006 may actually be higher in numerical value than the same data labeled as 2007 from some other reference source (nothing ever goes down in the energy world). As you go through the sections of this handbook and begin to get a feel for the global energy flow, you will see that the exact date of a particular set of data is secondary to the understanding of the energy-flow process itself.

Where data sets were available over a longer time period (like from 1980 to 2004), but did not extend to 2008, the approach was to graphically extrapolate to 2008 and beyond as necessary. In cases where data was available only for a particular year (2002, for example), growth-rate escalation factors were used to extrapolate to the 2007-08 time frame. Examples of data extrapolation and escalation to future years are found in Section 11.

## Section 1: What Is This Handbook All About?

Here is a summary of the more important ground rules and definitions used throughout this handbook.

| Table 1-2: Summary of Ground Rules and Definitions |
| --- |
| World energy perspective – mostly from EIA International Outlooks. |
| Reference time frame – the years 2007-2008. |
| Energy data is on a per-year basis – no energy per day, per week, per month. |
| Energy data is normalized to BTU (or quadrillion BTU) units. |
| Energy produced or generated is termed "Prime Source Energy" (PSE). |
| Energy consumed, delivered or used is "End-Use Consumption" (EUC). |
| Fossil prime energy sources are: oil, coal and natural gas. |
| Fossil-fuel derivatives are: gasoline, diesel fuel, fuel oil, hydrogen, propane. |
| Conventional energy sources are: fossil fuels plus hydroelectric and nuclear. |
| Alternative energy sources are: biomass, ethanol/biofuel, wind, solar, and geothermal. |

**The Global Energy Flow Chain – A Preview**

To interpret any energy information – text, tables, graphs or figures – you will need to know where you are in the *Global Energy Flow Chain*. The chain begins with the production of energy from the various prime sources and ends with its delivery to end-use consumers. The energy chain has multiple steps and paths before reaching the end-use stage. The energy flow-chain diagrams given in Section 3 will guide you through the flow process. Problems can often arise in interpreting unfamiliar energy information when it is not clear where you are in the flow of things. The energy-flow chains will help you to interpret almost any global-energy information that you may come across.

## Section 1: What Is This Handbook All About?

The energy-flow chain will be described in more detail (maybe too much) in Section 3 (Where Does All the Energy Go?). For now the chain can be thought of as a major river with many input tributaries. The river then divides into smaller rivers and streams. These may then recombine before reaching their final destinations. Caution, the *river of global energy* has unique characteristics that no other river has: it is always rising, it is never stagnant and it never recedes; it is relentless in its course. This may sound overly dramatic, but as you will see, that's the way it has been.

In this simple river analogy the input tributaries are the global prime energy sources. The final destinations for this river are almost infinite in their diversity. These destinations are initially grouped into the industrial, transportation, residential and commercial sectors. These sectors in turn are divided into numerous sub-sectors which will be discussed in Section 3.

When going down this river of energy, it is useful to have a map and compass to figure out where you have been and where you are going. The *Global Energy Flow Chain* will provide you with this map and compass.

### Summary of Section 1

The interpretation and understanding of global energy information is made difficult by the use of many and varied energy units and terms.

In this handbook all forms of global energy will be described in a consistent of set of energy units, the BTU.

Prime Source Energy (PSE) and End-Use Consumption (EUC) are the two most important terms used in describing the flow of global energy.

Understanding the *Global Energy Flow Chain* is onc of the main objectives of this handbook.

# Section 2

# The Sources of Global Energy

There are literally hundreds of studies and documents to be found on the Internet dealing with the sources and production of global energy. Most cover very specific topics like "Solar Energy Generation in Tanzania" or "Projections of Gasoline Production in the USA." These studies, as useful as they are to professionals in the energy field, are of little value to the general public simply because of the difficulty of putting the facts and figures into any meaningful global perspective.

## Energy Information – A Mixed Bag

This handbook presents the production and consumption of global energy in terms of a common and consistent set of energy units. The *British Thermal Unit* (BTU) is used throughout this work. Don't stop reading yet, at this point you don't even have to know what a BTU is. Nor will you have to understand megajoules, kilowatts, Mtoes, therms or any other mystifying energy units to find this work useful.

Petroleum is the world's number-one source of energy (as if you didn't already know), but how do the other energy sources rank? How about nuclear or hydroelectric power? Are these increasing in global terms? How about the alternative energy sources; solar, biofuels, wind and geothermal? You may have heard that wind energy capacity has increased by 1,000 megawatts in your state, or that ethanol production has reached 15 x $10^9$ liters per year in Brazil. You may ask, so what? What do these things mean from a global perspective? How much energy is this compared to the world's total energy production? This handbook will help you answer questions such as these.

## Section 2: The Sources of Global Energy

### The Energy Lineup

Figure 2-1 shows world prime source energy production in terms of percentage of the prime-fuel types. In Figure 2-2 we will look these in terms of their everyday working units (barrels, gallons, kilowatts, etc.). These will then these will be converted to a consistent set of energy units – the BTU.

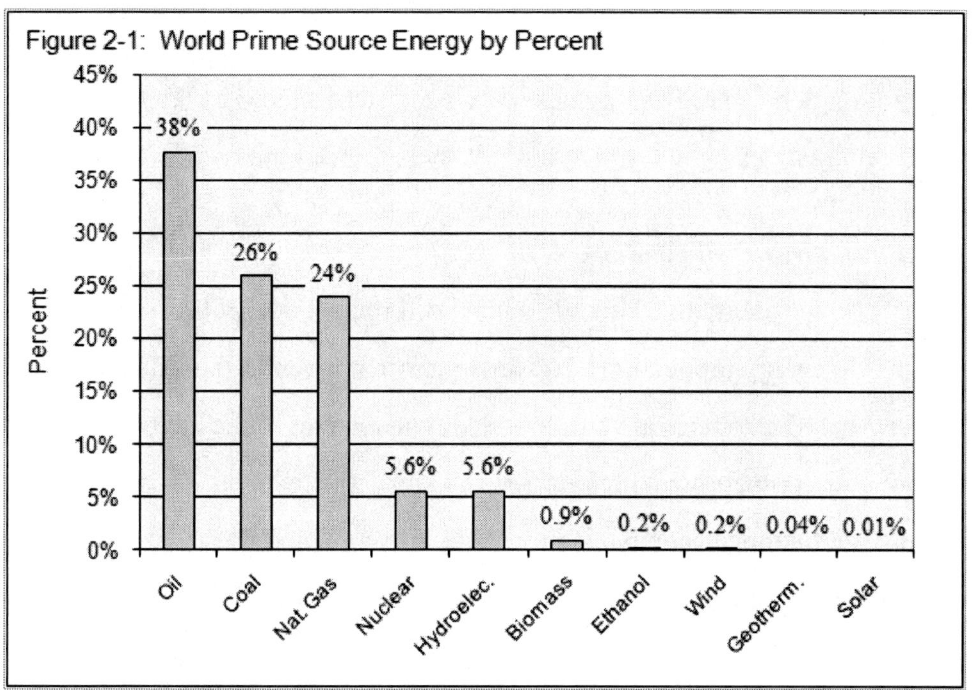


The big hitters in the energy world are the fossil fuels . . . for now.

Figure 2-1 tells us that the predominant percentage of global energy comes from fossil fuels. These supply nearly ninety percent of the world's energy. Next in line are nuclear and hydroelectric energy at about six percent each. These are followed by the alternative energy sources. The alternative sources have a combined percentage of somewhere between one and two percent of world energy production depending on the source of information and the definition of the term "alternative."

## Section 2: The Sources of Global Energy

### The Alternatives

What is most surprising from Figure 2-1 is the low percentage of global energy represented by the alternative sources. From all the articles, TV documentaries and general media attention given to alternative energy over the past several decades (since the 1970s oil crisis), you would think that the percentage values would be much higher. The threats of global warming and impending global energy shortages (take your pick) should be reason enough for the alternatives to be farther along. Why is this not the case?

*The Alternatives Trail the Pack*

Why do the alternative energy sources still remain such a low percentage of total world energy production? The answers often given to this question are: 1) fossil-based resources are still too readily available, 2) the alternatives are not yet cost-competitive with the fossils, 3) inadequate levels of government and private funding accompanied by a lack of vision and 4) the vested interests of oil and other energy-related corporations and their desire to maintain the status quo.

*The Fossils – Very Tough Competition*

There may be a little truth in the above, but the real answer has much more to do with the unique characteristics and capabilities of the fossil fuels:

***High energy-concentration – Greatest bang for the buck.***
***Transportability – Can be delivered to virtually anywhere in the world.***
***Storage capability – Can be stored anywhere in the world, indefinitely.***
***Full-time worker – No down-time, always ready to perform.***
***Readily Available – Even with the threat of dwindling supplies.***

## Section 2: The Sources of Global Energy

### The Energy Heavyweights

What a lineup! Fossil fuels are the heavyweight champs of world energy. Later we will compare each of the energy sources in terms of their specific energy or energy storage capacity. This will give you an even better feel for why the fossil fuels remain on top. For now just think about how much energy there is in a gallon can of gasoline compared to the energy in a common auto battery. Both are of about the same approximate size, but the battery is much heavier. The can of gasoline contains 125,000 BTU of energy; the battery contains only about 1,600 BTU. This is an incredible difference in energy content of 78 to 1. Yes, I know there are more efficient batteries than the common auto battery, but even these don't come anywhere near the energy-storing capacity of gasoline (see Section 6, Table 6-3b).

### It's All About BTU

Now it is time to look at global energy production in terms of a consistent set of energy units. Because global energy consumption values are of such high magnitude, they are commonly expressed as quadrillion BTU or QBTU. What the hell is a QBTU you ask? It's the next numerical-descriptive term after a trillion. You know, a million = $10^6$, a billion = $10^9$, a trillion = $10^{12}$ and therefore a quadrillion must be a number to the $10^{15}$ power.

#### *What Is a BTU?*

At this point an explanation of the BTU is usually given in works like this. All of the numerous unit systems for quantifying energy – such as calories, ergs, kilowatt-hours, joules or Mtoes, etc. – can be converted into BTU. The BTU can be defined as the amount of energy required to increase the temperature of a given quantity of water so many degrees. I am not going to explain this in numerical terms. Look it up if necessary. The important things you need to know about the BTU and energy are:

## Section 2: The Sources of Global Energy

1. One barrel of crude oil contains a lot of BTU (5.8 million).

2. The world's energy supply from all sources for 2007-2008 is 490 QBTU (plus or minus a few percent, depending on who is keeping score).

3. Fossil fuels make up over 87 percent of these 490 QBTU.

4. Approximately 189 QBTU (39%) of the 490 QBTU goes into the generation of the world's electricity supply.

5. The remaining 301 QBTU (61%) is used directly as gasoline, natural gas, coal, or as petroleum derivatives such as diesel and aviation fuel, heating oil and also as source materials for the petrochemical industry.

*Converting to BTU*

When you read about energy sources and derivatives, they are usually described using their "working units" – oil in barrels, gasoline in gallons, etc. The conversion of working units to BTU is straightforward for the fossil fuels. It is a little more involved with electricity, as you will see later. Here are some useful factors for converting from common working units to BTU;

Barrels of oil times $5.86 \times 10^6$ = BTU

Cubic feet of natural gas times 1,030 = BTU

Short tons of coal times $2.07 \times 10^7$ = BTU

One gallon of gasoline times $1.25 \times 10^5$ = BTU

One gallon of ethanol times $7.6 \times 10^4$ = BTU

Megajoules times 947.8 = BTU

Mtoes times $3.9 \times 10^{13}$ = BTU

## Section 2: The Sources of Global Energy

Of course when you actually need to convert something to BTU, it often won't be this simple. Natural gas may be given in "therms," coal may be in "metric tonnes." You may have to make two or three intermediate conversions before you finally get to BTU. If you really need to play the energy-conversion game, there are numerous Websites that give you automated conversion factors for almost anything to anything. One of the most useful of these is AllConversions.com. Another useful energy conversion site is found under the EIA Kids series, (eia.doe.gov/kids/energyfacts/science/energycalculator).

*Electricity Conversion to BTU*

The working units for electrical power are kilowatts or megawatts; these are not directly convertible to BTU. They must first be multiplied by some known or assumed number of effective operating hours to obtain kilowatt-hours or megawatt-hours (energy units) which can then be converted to BTU. The *Energy Reference Tables* in Section 4 (Energy in Everything) give the yearly operating hour factors assumed in this work for the electrical power sources that are initially expressed in megawatt or kilowatt units.

Here are some useful electrical energy conversion factors:

Kilowatt-hours times 3,412 = BTU

Megawatt-hours times 3.412 x $10^6$ = BTU

Kilowatts times effective op. hours times 3,412 = BTU

### Global Energy in BTU

Table 2-1 gives the world's prime sources of energy in their common working units. These values are converted to BTU and then finally to QBTU. The converting of all energy values to a consistent set of QBTU values allows for the making of rapid "apples-to-apples" comparisons between each of the energy sources, as you will soon discover.

## Section 2: The Sources of Global Energy

**Table 2-1: World Prime Source Energy Production – 2008**

| Prime Source | World Percent | PSE in Working Units | | World PSE BTU | World PSE QBTU |
|---|---|---|---|---|---|
| Oil | 38% | 3.16E+10 | Barrels/yr. | 1.85E+17 | 185 |
| Coal | 26% | 6.35E+09 | Short tons/yr. | 1.27E+17 | 127 |
| Nat. Gas | 24% | 1.14E+14 | Cubic feet/yr. | 1.17E+17 | 117 |
| Nuclear | 5.6% | 3.61E+05 | Megawatts | 2.7E+16 | 27 |
| Hydroelec. | 5.6% | 9.00E+05 | Megawatts | 2.7E+16 | 27 |
| Biomass | 0.9% | 4.30E+15 | BTU/yr. | 4.30E+15 | 4.3 |
| Ethanol | 0.2% | 1.33E+10 | Gallons/yr. | 1.0E+15 | 1.0 |
| Wind | 0.2% | 9.00E+04 | Megawatts | 9.3E+14 | 0.9 |
| Geotherm. | 0.04% | 1.00E+04 | Megawatts | 2.1E+14 | 0.2 |
| Solar | 0.01% | 9.00E+03 | Megawatts | 6.7E+13 | 0.07 |
| **Total >** | **100%** | ----- | ----- | **4.90E+17** | **490** |
| *All Fossil* | 87.5% | ----- | ----- | 4.29E+17 | *429* |
| *All Non-Fossil* | 12.5% | ----- | ----- | 6.11E+16 | *61.1* |
| *All Alternatives** | *1.33%* | ----- | ----- | *6.52E+15* | *6.5* |

** Biomass, wind, geothermal, ethanol and solar*

This energy table will help you make sense of the many mixed and confusing units for energy production (PSE) that you usually run across in the media or technical reports.

The EIA usually gives world fossil-fuel data in QBTU, but often gives electricity data in terms of kilowatts or megawatts. Sometimes, however, the EIA gives electricity in kilowatt-hours or megawatt-hours. One must be careful not to get the two confused. In Table 2-1 the working units for the electricity-producing power sources (Megawatts) have been converted to BTU using data available from the Internet or were estimated using the effective yearly operational hour factors as given in the *Energy Reference Tables* of Section 4.

## Section 2: The Sources of Global Energy

Figure 2-2 is a bar chart showing the world's prime source energy production in QBTU. Fossil fuels provide an overwhelming 429 QBTU of the 490 QBTU of world energy. Nuclear and hydroelectric energy each provide about 27 QBTU. All of the alternative energy sources combined provide a little over 6 QBTU. This bar chart makes the small contribution of the alternative sources to the global energy supply all too apparent.

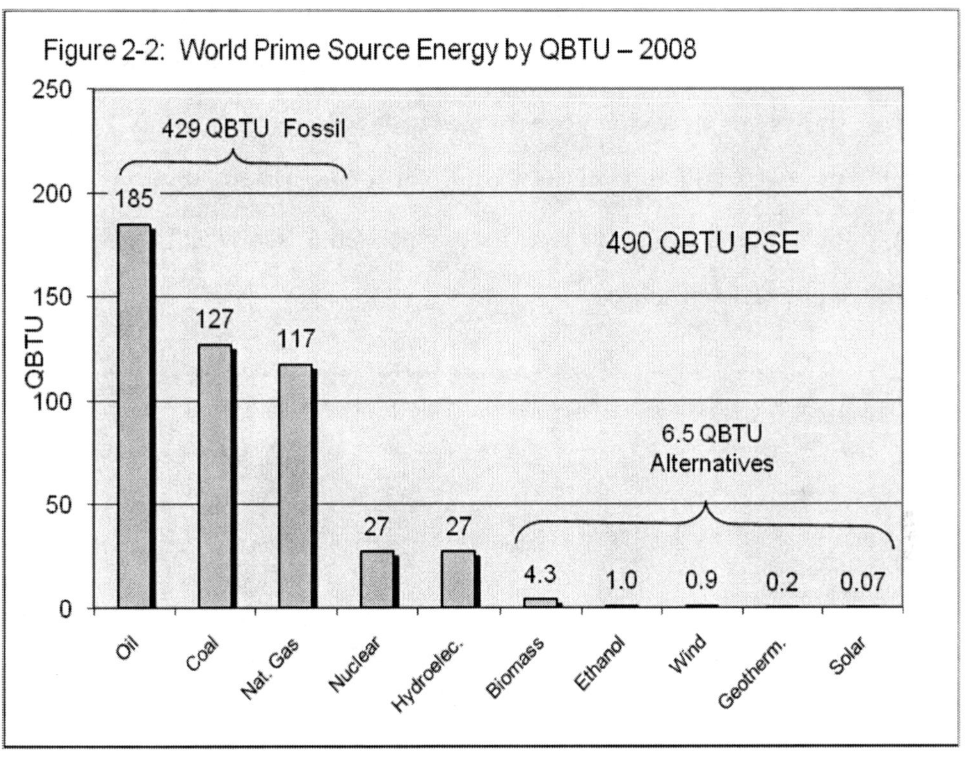

Figure 2-2: World Prime Source Energy by QBTU – 2008

World energy production is dominated by the fossil fuels. The alternative energy sources, while steadily growing in their own right, have a difficult time keeping up with the ever increasing demand for fossil fuel.

## Section 2: The Sources of Global Energy

### Looking Back to 1980

Figure 2-3 tells an interesting story. Oil use shows a significant downward slope associated with the Iraq-Iran War and resultant energy crisis of the early 1980s. Oil then shows a consistent growth that is predicted to continue through the 2030s. Coal showed a consistent growth rate during most of the 1980s, but experienced a flat period during the 1990s. Since then, coal production has been on a continual upswing. Nuclear energy showed its greatest growth during the 1980s, but has leveled off to a uniform growth rate since then. Hydroelectric growth is best described as slow but steady over the entire twenty-five years. The chart shows an improving growth rate for the alternative energy sources from now through 2030, but still remaining far behind the ever-growing consumption of fossil fuels. A closer look is needed to see what is really going on with the alternatives sources.

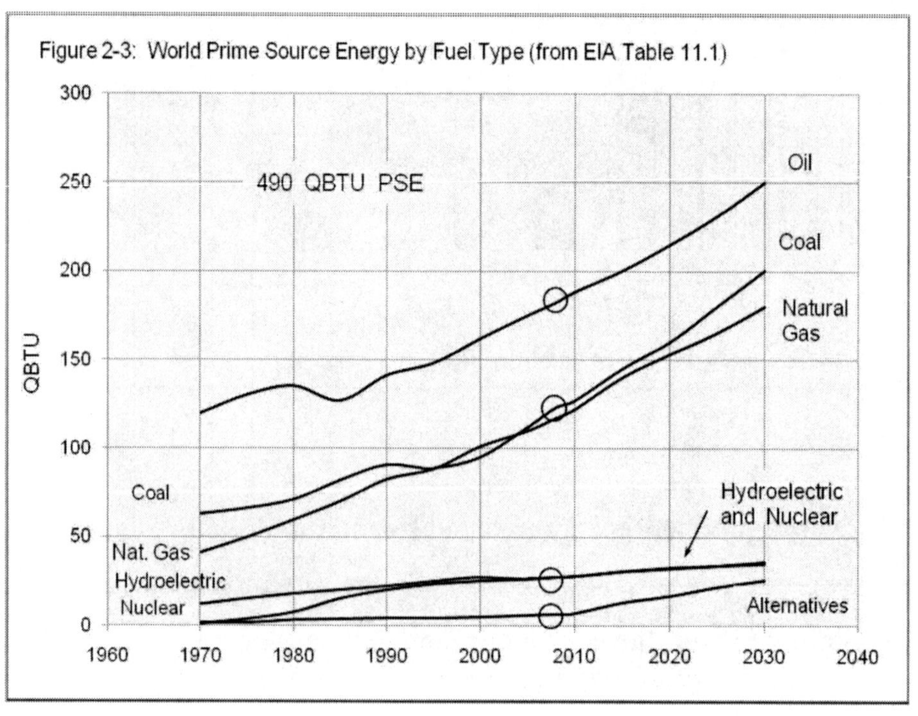

## Section 2: The Sources of Global Energy

### Alternative Energy Growth

When viewed independently from the fossil fuels, the growth of alternative energy looks a lot better as shown in Figures 2-4 and 2-5. The alternative energy sources have shown a steady growth rate since 1970, but a much higher (skyrocketing) rate is predicted through 2030 and beyond. The recent United Nations report on global warming (2007) and the responses by the U.S. and the international community will hopefully ensure the occurrence of this high growth rate as predicted by the EIA.

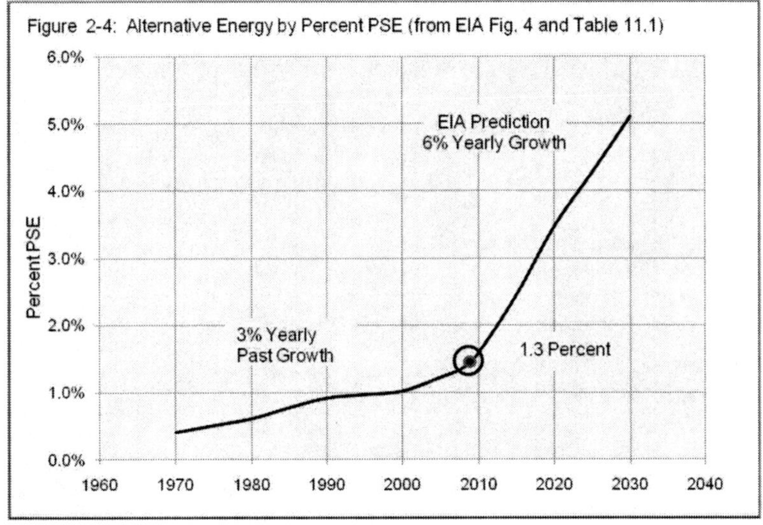


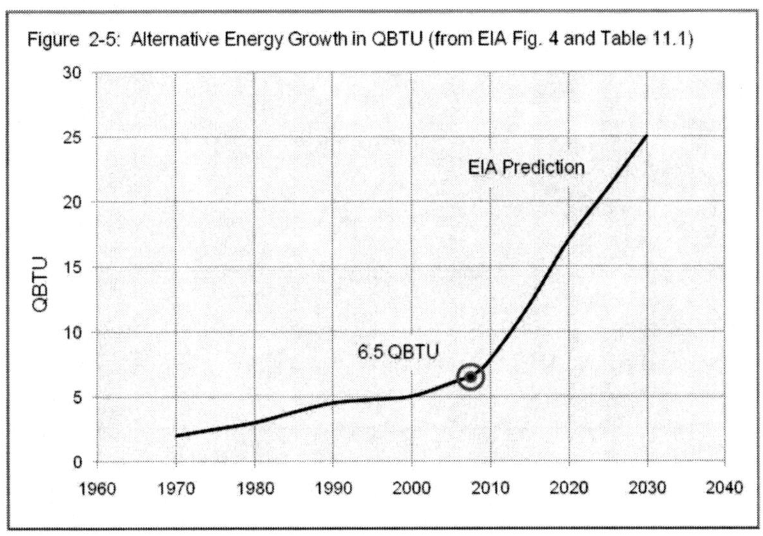

## Section 2: The Sources of Global Energy

### Global Energy Growth

We have seen that global energy production for the 2007-2008 period is 490 QBTU. What is expected in future years? How does this compare with past energy growth? Figure 2-6 gives historic global energy production from 1970 through 2008, along with EIA projections to the year 2030. The predictions say global energy production will continue to increase at about two percent yearly through 2030 and beyond. The energy growth rate shown here is based on EIA predictions using their pragmatic "2.0 percent" growth scenario. It should be noted that the EIA (as well as other energy information sources) also gives energy forecasts based on several other growth scenarios, many of which are more optimistic for reducing fossil-fuel consumption and for the growth of the alternative sources (see Figure 2-7).

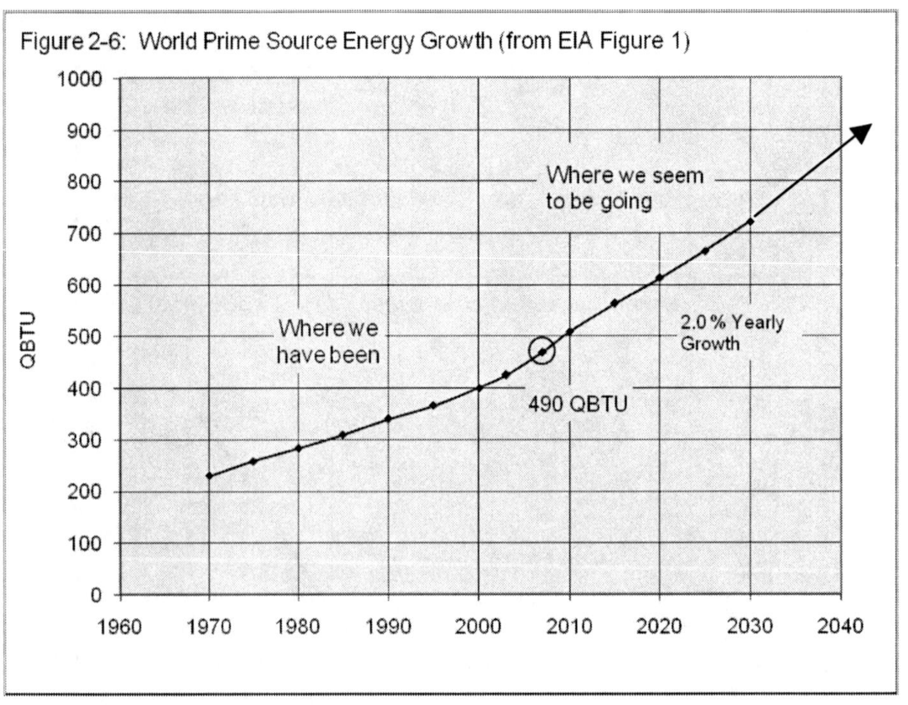

Figure 2-6: World Prime Source Energy Growth (from EIA Figure 1)

## Section 2: The Sources of Global Energy

Figure 2-7: EIA World Energy Growth Scenarios (EIA Fig. 16)

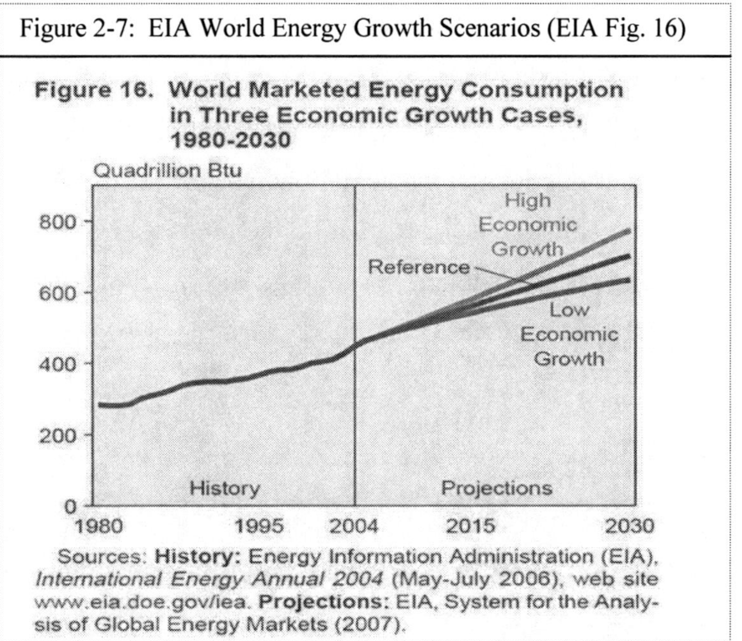


### Summary of Section 2

World energy demand is expected to continue increasing at the rate of about two percent yearly through 2030.

Petroleum is expected to remain the major source of global energy despite impending shortages, increasing costs and being a major contributor to greenhouse gasses and global warming.

Coal and natural gas production will increase at even higher growth rates than in past years – this due to the increased demand for electricity and also the rising cost of petroleum.

The alternative energy sources will continue to increase in numerical value (QBTU), but will remain a relatively small percentage of total world energy production through 2030.

Hydroelectric and nuclear energy production growth rates will continue at modest levels compared with fossil-fuel growth rates.

# Section 3

# Where Does All the Energy Go?

In this section the flow of prime source energy will be followed to its end use. The *Global Energy Flow Chain*, briefly covered in Section 1, will be described here in more detail. Once you obtain a feel for the flow chain, you should find it very useful for understanding the remainder of this work, as well as for interpreting the landslide of new energy information bound to come your way as fuel prices increase and global warming issues become more heated.

## The Energy Flow Chain

Before looking at the actual energy values that comprise the energy-flow chain, we need to define some terms and relationships. You may be wondering why it's just not called the energy-flow diagram. It is referred to as a chain because each of the elements are linked (chained) to each other. Once any three elements in the chain are defined, the remaining elements are set and can be determined by calculation. All energy elements must balance out to the total amount of world energy produced for that year. If one energy element is found to be larger than first thought, then some another element must therefore be smaller. This will be made clear in the following discussion, which shows how the elements of the flow chain are mathematically related to each other.

### *Definition of Flow Chain Terms*

In order to think about the elements of the energy-flow chain, some terms need to be defined:

1. Total Prime Source Energy will be represented by the term PSE(t)
2. Total End-Use Consumption will be represented as EUC(t)
3. Prime Source Energy used for electricity generation as PSE(e)
4. Electricity End-Use Consumption as EUC(e)
5. Prime Source Energy directly consumed (non-electric) as EUC(d)
6. Prime Source Energy lost will be represented as PSE(l)

## Section 3: Where Does All the Energy Go?

Internet information on PSE(t), EUC(t) and EUC(e) are usually the easiest to locate and identify. Data on the remaining terms are more difficult to find or at least more difficult to identify. Often you will not be sure of the category of your data. Cross-checking some specific values with other known data sets may help you determine where your data lies within the flow chain.

The six energy elements in the flow chain are related to each other by these simple equations:

$$\text{EUC(d)} = \text{EUC(t)} - \text{EUC(e)}$$

$$\text{PSE(e)} = \text{PSE(t)} - \text{EUC(d)}$$

$$\text{PSE(l)} = \text{PSE(e)} - \text{EUC(e)}$$

$$\text{PSE(t)} = \text{EUC(t)} + \text{PSE(l)}$$

No need to memorize these equations. The relationship between the elements of the flow chain will be shown clearly in the series of diagrams which follow.

*The Energy Flow Chain*

Figure 3-1 is a diagram of the flow chain at the first level. Here the basic terms of the chain (as given above) are indicated but without the numerical values of energy associated with them. These will be included as we move to the remaining diagrams.

Phase 1 – Prime source energy, PSE(t), first separates into the categories of direct energy consumption, EUC(d), and into the energy used for thc generation of global electricity, PSE(e).

## Section 3: Where Does All the Energy Go?

Phase 2 – The electricity generation process results in a major split in the flow chain. The conversion of fossil fuels into heat and steam to produce electricity is accompanied with significant energy losses. Further energy losses occur from the conditioning and transmission of the electricity to end-use consumers. Both of these losses are combined in the PSE(l) term. The amount of electrical energy reaching the end-use consumption phase is represented by the EUC(e) term.

Phase 3 – Direct-use energy, EUC(d), and end-use electrical energy, EUC(e), recombine to make up total world consumed energy, EUC(t). The amount of energy, in QBTU, for each step of the flow chain will be given in the Figure 3-2, next page.

**Figure 3-1: The Energy Flow Chain**

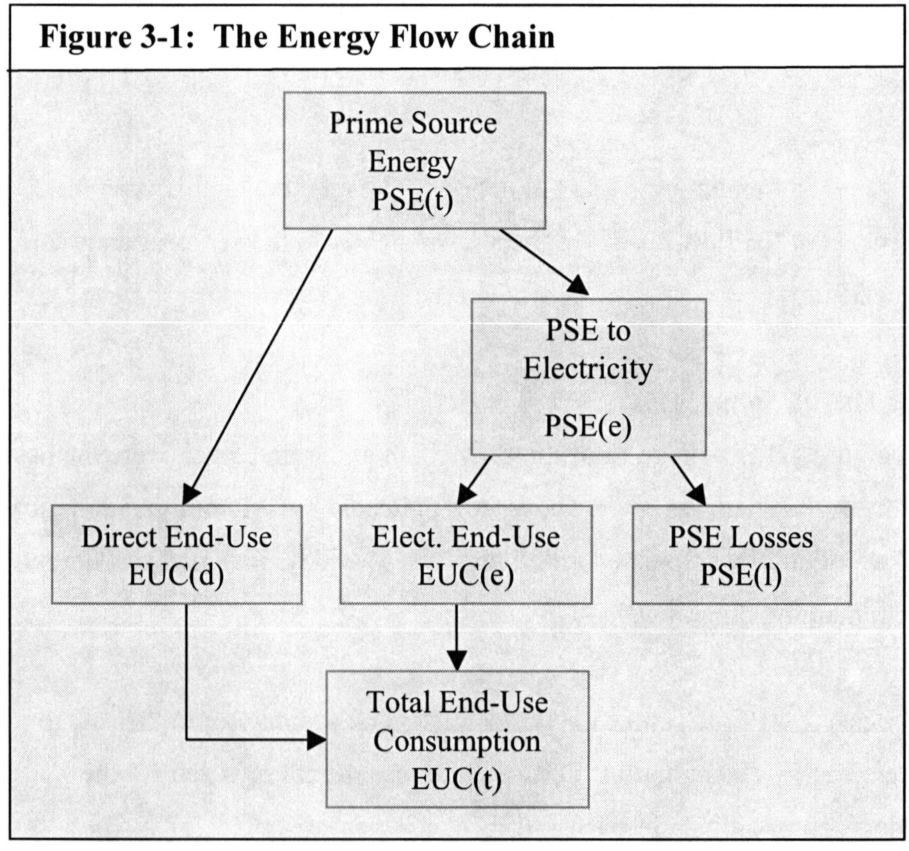

## Section 3: Where Does All the Energy Go?

*The Energy Flow Chain with QBTU Values*

To see the actual amount of energy involved, we need to insert QBTU energy values into the flow chain. This is done in Figure 3-2. The figure shows that of the initial 490 QBTU, 301 QBTU is directly consumed in the form of gasoline, fuel oil, heating oil, natural gas, coal and petrochemical products. The remaining 189 QBTU goes for electricity generation. Of this, approximately 126 QBTU is lost during energy conversion and subsequent transmission to consumers. This leaves 63 QBTU of electrical energy actually being delivered to end-use consumers. This electrical energy, when combined with the 301 QBTU of direct-use energy, results in a total of 364 QBTU of end-use consumed energy, EUC(t).

**Figure 3-2: Energy Flow Chain in QBTU**

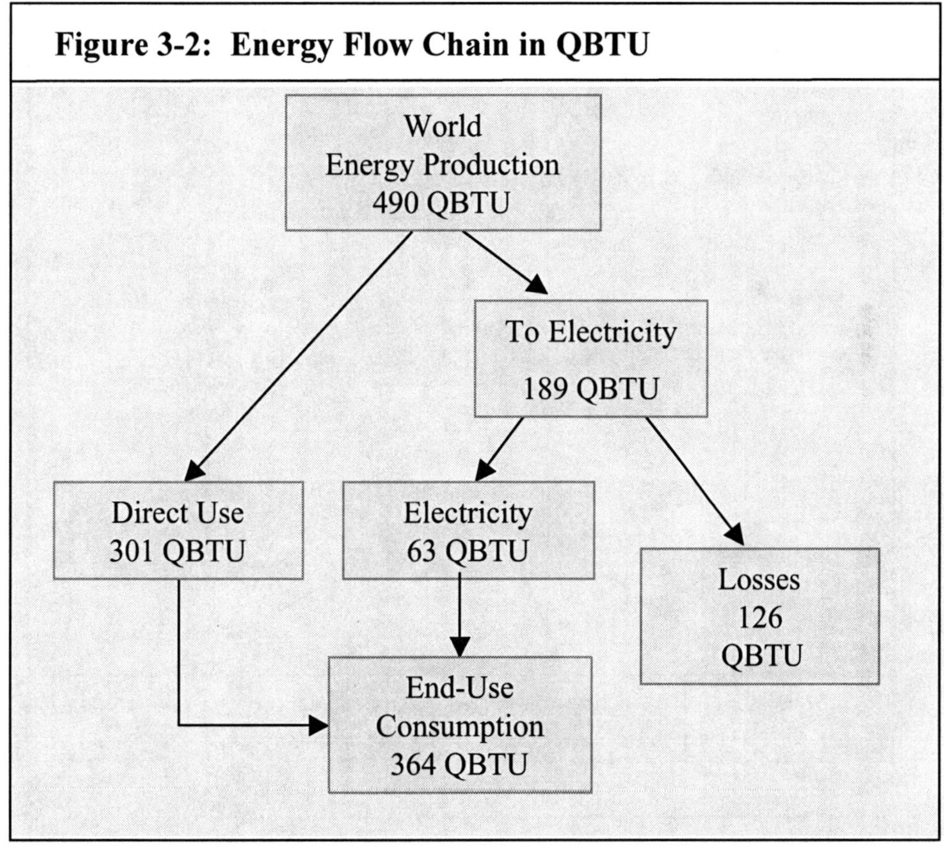

## Section 3: Where Does All the Energy Go?

### Energy Flow to the Sectors

To understand where energy goes next, we need to look at Figure 3-3. This figure shows the amount of energy reaching the four major energy consumption sectors – industrial, transportation, residential and commercial – as commonly defined in the field of energy information and by the EIA. These major sectors, in turn, are broken down into numerous sub-sectors. These will be looked at further when we get to the last level of the energy-flow chain (Figure 3-7).

**Figure 3-3: Energy Flow Chain to the Sectors**

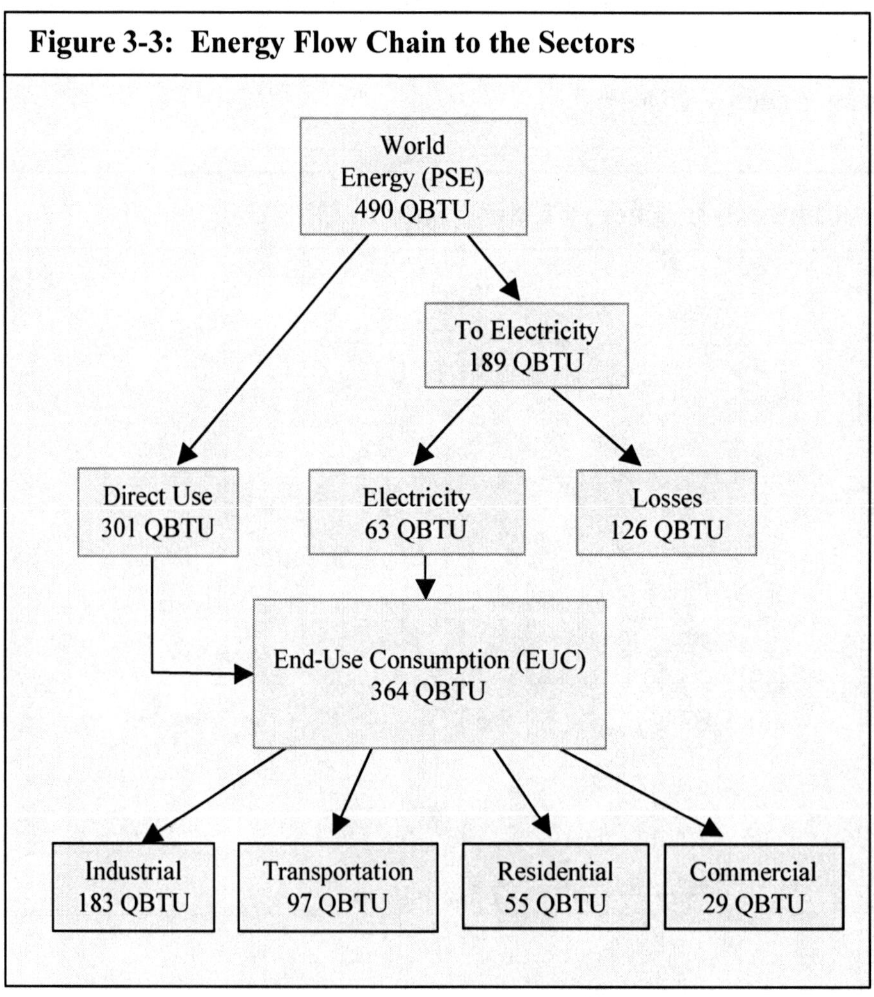

## Section 3: Where Does All the Energy Go?

*Introducing the Sectors*

There are many possible ways in which the consumption sectors and sub-sectors can be defined and organized. In this work the sub-sectors are grouped into the categories listed in Table 3-1. These groupings were mainly chosen to make certain points about energy consumption. Other authors and sources may use other groupings.

In the *Residential Sector* for example, clothes dryers are listed as a separate item – usually they are combined with the other home appliances. Since dryers are often accused of being energy wasters, separating them allows us to see how much energy they really use. The appliances-plus category includes anything that is plugged-in – kitchen items, washing machines, electronic devices, garden and workshop tools, swimming-pool pumps and so on. If you are interested in finding out about the energy consumption for a specific household item, your utilities company probably has a Website that will help you with this.

The *Transportation Sector* is divided into the *Motor Vehicle* and the *Other Transportation* sub-sectors. The motor vehicle sub-sector is responsible for 32 percent of all petroleum consumption, but only 14 percent of all fossil-fuel consumption. This may help dispel the notion that we can significantly reduce global warming by simply driving more energy-efficient cars. In Section 9 the gasoline savings from improved fuel economy is played off against the increasing number of autos in the world. Guess who wins in this game.

The *Commercial Sector* is broken into the *business/finance* and the *services* sub-sectors. The energy consumed in this sector – by commerce, government, defense, health, education, entertainment and communications – serves us daily, but we are usually unaware of its presence.

## Section 3: Where Does All the Energy Go?

The *Industrial Sector* is broken into the three sub-sectors: *Material/Resources*, *Manufacturing/Construction* and *Agriculture.* Consumer spending and energy consumption is closely tied to the industrial sector – either directly through the energy used for the production of goods or indirectly through the energy for extracting and processing of natural resources.

| **Table 3-1: The End-Use Energy Consumption Sectors** |
|---|
| **The Industrial Sector** |
| Materials and Resources (metals, chemicals, refining, wood, paper). |
| Manufacturing and Construction (anything produced or built). |
| Agriculture (growing, irrigation, nutrients, processing). |
| **The Transportation Sector** |
| Motor Vehicles (autos, light trucks, other gasoline vehicles). |
| Other Transport (air, rail, trucking, shipping, public transit). |
| **The Residential Sector** |
| Space Heating (aka room heating, electric or fuel). |
| Air Conditioning (as in "set at 80 degrees"). |
| Appliances-plus (anything that plugs-in). |
| Home Lighting (incandescent or as use "fluorescent"). |
| Water Heating (could be heating fuel or electric). |
| Clothes Dryers (as "use your clothesline"). |
| **The Commercial Sector** |
| Business and Finance (banks, insurance, wholesale, retail). |
| Services (health, education, government, entertainment, communications). |

## Section 3: Where Does All the Energy Go?

*The Industrial Sector*

Fifty percent of all global energy consumption occurs in the *Industrial Sector.* The *Manufacturing/Construction* sub-sector of the industrial sector covers the production of paper clips to computers to appliances to automobiles to Boeing 747s, and anything else you can think of that is produced in the world. Construction includes everything from houses to highways, from buildings to bridges, and from water systems to mega-hydroelectric dams. The *Materials and Resources* sub-sector includes the production of energy itself, as in oil refining and the manufacturing of the thousands of chemical and plastic materials used throughout the world. Paper and related products are one of the biggest energy-using areas in this sub-sector, as is the making of steel and the many other industrial metals. This sub-sector also includes the energy to find, develop and process the numerous natural resources (forest products, water resources, minerals, etc.) required to sustain the industrial sector. *Agriculture* is a sub-sector of the industrial sector, although because of its importance, could easily be a major sector unto itself.

*The Agriculture Sub-Sector*

This sub-sector covers all the activities involved in the growing, harvesting, initial processing and storing of the world's food supply for the six billion plus of us – a tall order to say the least. There seems to be no consensus on what this sub-sector encompasses. Estimates range from 5 to 20 percent of total world energy depending on how far up the global food chain you go. If commercial food processing, distribution, wholesale and retail activities were included it could be over 20 percent of world energy. A nominal value of 10 percent of global EUC was selected here to represent the agricultural sub-sector. This equates to about 8 percent of global PSE. This covers the energy for growing and initial processing. Activities beyond this would be in the other sectors; factory processing would be in the industrial sector, distribution in the transportation sector and wholesale/retail activities would be in the commercial sector.

## Section 3: Where Does All the Energy Go?

*The Transportation Sector*

The *Transportation Sector* needs little explanation. At 27 percent of world end-use energy consumption, this sector literally makes the world go around. It includes cars, trucks, planes, ships and anything else that moves people and things from place to place. It should be noted that petroleum is the primary source of energy for this sector. Coal, natural gas and electricity are relatively small contributors here. Their percentages can hardly be seen on the charts. The transportation sector will be discussed in terms of two sub-categories: the *Motor Vehicle* sub-sector, which includes our cars and the *Other Transportation* sub-sector, which includes the trucking, airline, railway and shipping industries.

*The Residential Sector*

The *Residential Sector* accounts for about 15 percent of the world's end-use energy. This sector and the motor vehicle sub-sector are the most familiar to us on a day-to-day basis. The residential sector covers all of the energy we spend in our home including heating fuel as well as electricity. This sector is divided into the following sub-sectors: 1) *Heating*, 2) *Cooling,* 3) *Appliances,* 4*) Lighting*, 5) *Clothes Dryers*, and 6) *Water Heaters*. The breakdowns for residential energy found in other works will almost always be different from these. For example, home electronics are usually given as a separate item in U.S. residential energy breakdowns. Here they are rolled into the general appliance category.

The residential sub-sectors given here were chosen to point out several of those items often found in typical Sunday-supplement articles dispensing tips on saving energy at home. A typical article might be entitled "Twelve Tips for Saving Energy at Home," where Tip # 7 might be "use a clothesline to dry your wash, not your clothes dryer."

## Section 3: Where Does All the Energy Go?

Because of the difficulty in obtaining consistent information on global residential energy, approximations were made using the more available U.S. residential data as the starting point. The intent is to give you a first-level idea of global residential energy use, not a detailed list of energy use to third decimal place. Data on residential energy consumption at the global level will always be very approximate considering the almost infinite variety of families and residences around the world.

*The Commercial Sector*

Last in line for energy consumption is the *Commercial Sector*. This sector accounts for about 8 percent of end-use energy consumption. The industrial sector can be thought of as the production of "things." The commercial sector can be thought of mainly as "services". This sector covers activities from mom-and-pop stores to department stores, from local banks to mega-financial corporations. The entertainment and communications industries are also included here. These come under the sub-sector labeled *Business and Finance* in the tables and graphs. The commercial sector also includes health, education, protection (police to military) and governmental operations. These are under the sub-sector labeled *Public Services* in the tables and graphs.

*You May Ask "And Where Is . . .?"*

With all of the ways in which global energy is expended, how can we be sure that everything is accounted for? Where, for example, is defense? Everyone knows that defense is a major government activity and that it must consume extraordinary amounts of energy at the global level. The answer is energy consumption for defense is spread between the four major sectors. Weapons are produced in the industrial sector; food, transportation and government operations are accounted for in the other appropriate sectors. By definition, all world energy consumption is made to fit somewhere within the four sectors.

## Section 3: Where Does All the Energy Go?

### End-Use Consumption by the Sectors

The energy consumed by the four major sectors is shown graphically in Figure 3-4. Note that the total consumed energy is only 364 QBTU, not the full 490 QBTU of total prime source energy we are familiar with from Section 2. This is because electrical energy losses are not accounted for in end-use consumption databases. In the remainder of this section we will be looking at how this 364 QBTU is consumed down to the various sub-sector levels. Here we begin to see why energy savings at the household level alone can only be a partial solution to our energy problems.

At the end of this section we will look at end-use consumption combined with energy losses. We will be seeing a somewhat different distribution of consumed energy (see Figure 3-9) than is shown here in Figure 3-4.

Figure 3-4: End-Use Consumption by the Sectors
(from EIA Figures 19, 21, 23, 25)

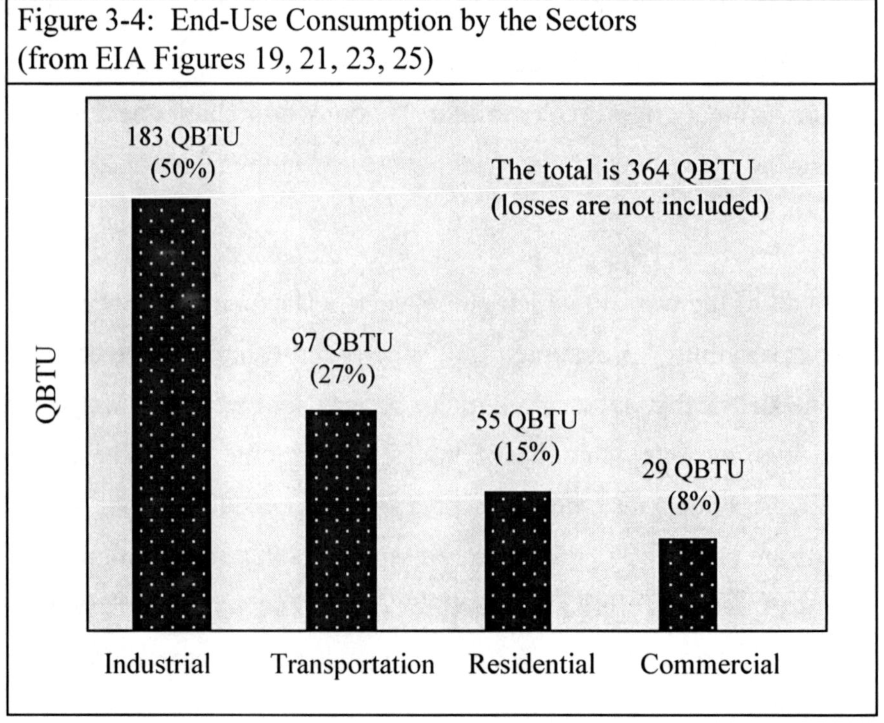

## Section 3: Where Does All the Energy Go?

*Sub-sector Consumption*

Table 3-2 shows how global energy is consumed down to the sub-sector level. The percentage values are given in terms of global end-use consumption (EUC). Note that energy losses are not included here. Often in tables that you may encounter, the sub-sector energy consumption values are given in terms of percentages within the sector itself. For example, residential heating can be correctly stated as 55 percent of the residential sector, or it can also be correctly stated as 8.3 percent of total global EUC, as is done in the table. This gets back to the "percentage of what" question and the need to look carefully at all energy data in terms of "percentage of what."

The table lists several energy-consuming items that you should be able to relate to in your daily life – clothes dryers, home lighting, appliances and motor vehicles.

**Table 3-2: World End-Use Consumption by Sector – 2008**

| Sector | Percent of World EUC | World EUC BTU | World EUC QBTU |
|---|---|---|---|
| **Industrial** | **50.4%** | **1.83E+17** | **183** |
| - Materials/Resources | 12.0% | 4.37E+16 | 44 |
| - Mfg./Construction | 28.4% | 1.03E+17 | 103 |
| - Agriculture | 10.0% | 3.64E+16 | 36 |
| **Transportation** | **26.7%** | **9.72E+16** | **97** |
| - Motor Vehicles | 16.5% | 6.01E+16 | 60 |
| - Other Transport | 10.2% | 3.71E+16 | 37 |
| **Residential** | **15.0%** | **5.46E+16** | **55** |
| - Space Heating | 8.3% | 3.04E+16 | 30 |
| - Air Conditioning | 1.0% | 3.64E+15 | 3.6 |
| - Appliances-plus* | 2.4% | 8.74E+15 | 8.7 |
| - Home Lighting | 1.1% | 4.00E+15 | 4.0 |
| - Water Heating | 1.8% | 6.55E+15 | 6.6 |
| - Clothes Dryers | 0.4% | 1.27E+15 | 1.3 |
| **Commercial** | **8.0%** | **2.91E+16** | **29** |
| - Business/Finance | 5.0% | 1.82E+16 | 18 |
| - Public Services | 3.0% | 1.09E+16 | 11 |
| **Totals** | **100%** | **3.643E+17** | **364** |

* Kitchen, electronics, tools, outdoor and gadgets, etc.

## Section 3: Where Does All the Energy Go?

*Sub-Sector Consumption – Percentage Bar Graph*

Figure 3-5 depicts global end-use energy consumption by sub-sectors in terms of percentage. This figure clearly shows who the heavy hitters are in world of energy consumption.

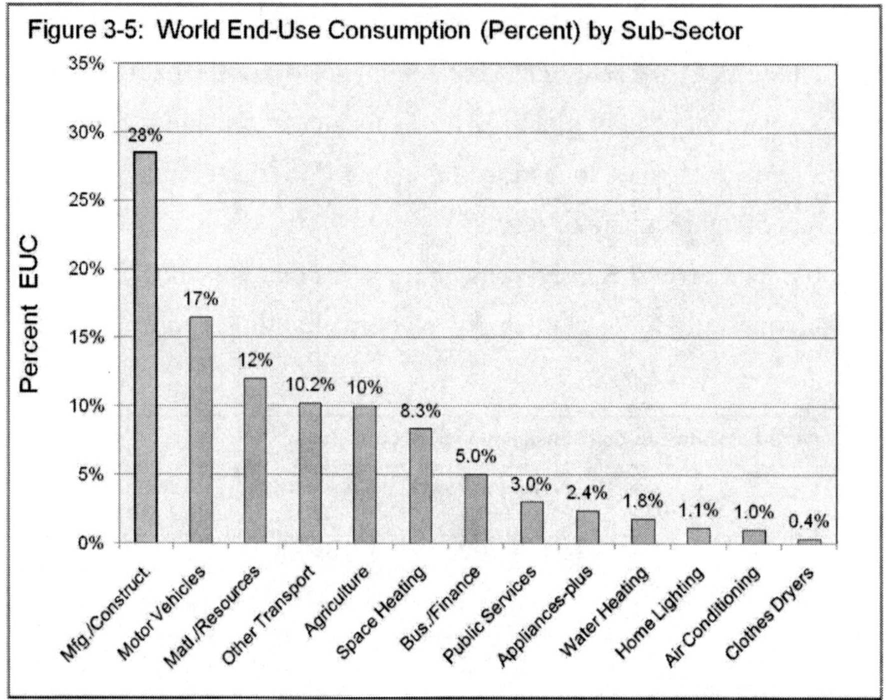


*Sub-Sector Consumption – QBTU Bar Graph*

Figure 3-6 (next page) graphically shows global energy consumption by QBTU down to the sub-sector level. This figure provides a comparative reference for the actual energy used within the major sectors and sub-sectors. In Section 9 (How to Use This Guide) you will see how this chart and others to follow can be useful for interpreting the often confusing information on energy-saving proposals that you routinely hear about.

## Section 3: Where Does All the Energy Go?

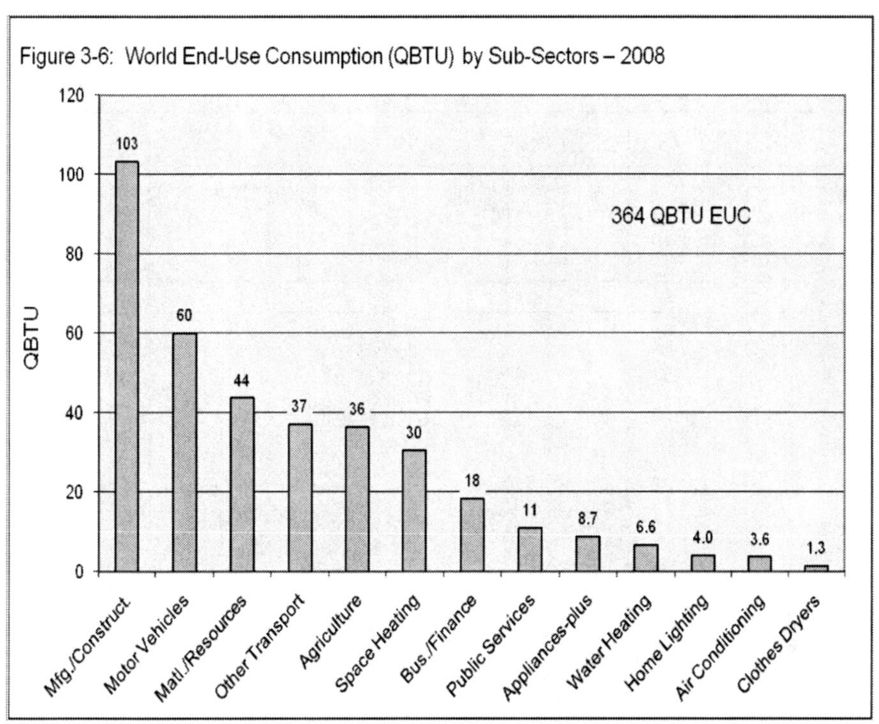


> This table shows global end-use energy consumption by QBTU for the major sectors and sub-sectors. Remember, end-use consumption does not include energy losses.

Manufacturing, construction and motor vehicles head the global energy consumption list. Residential space heating (room) is relatively high on the list – except for those who live in tropical regions, everyone else in the world, rich or poor, needs heat for their homes during the cold-weather seasons. The business/financial sub-sector comes in at less than 20 QBTU. The remaining sub-sectors are in the 10 QBTU or less range. The complete the *Global Energy Flow Chain* is given next in Figure 3-7.

## Section 3: Where Does All the Energy Go?

**Figure 3-7: The Global Energy Flow Chain – 2008**

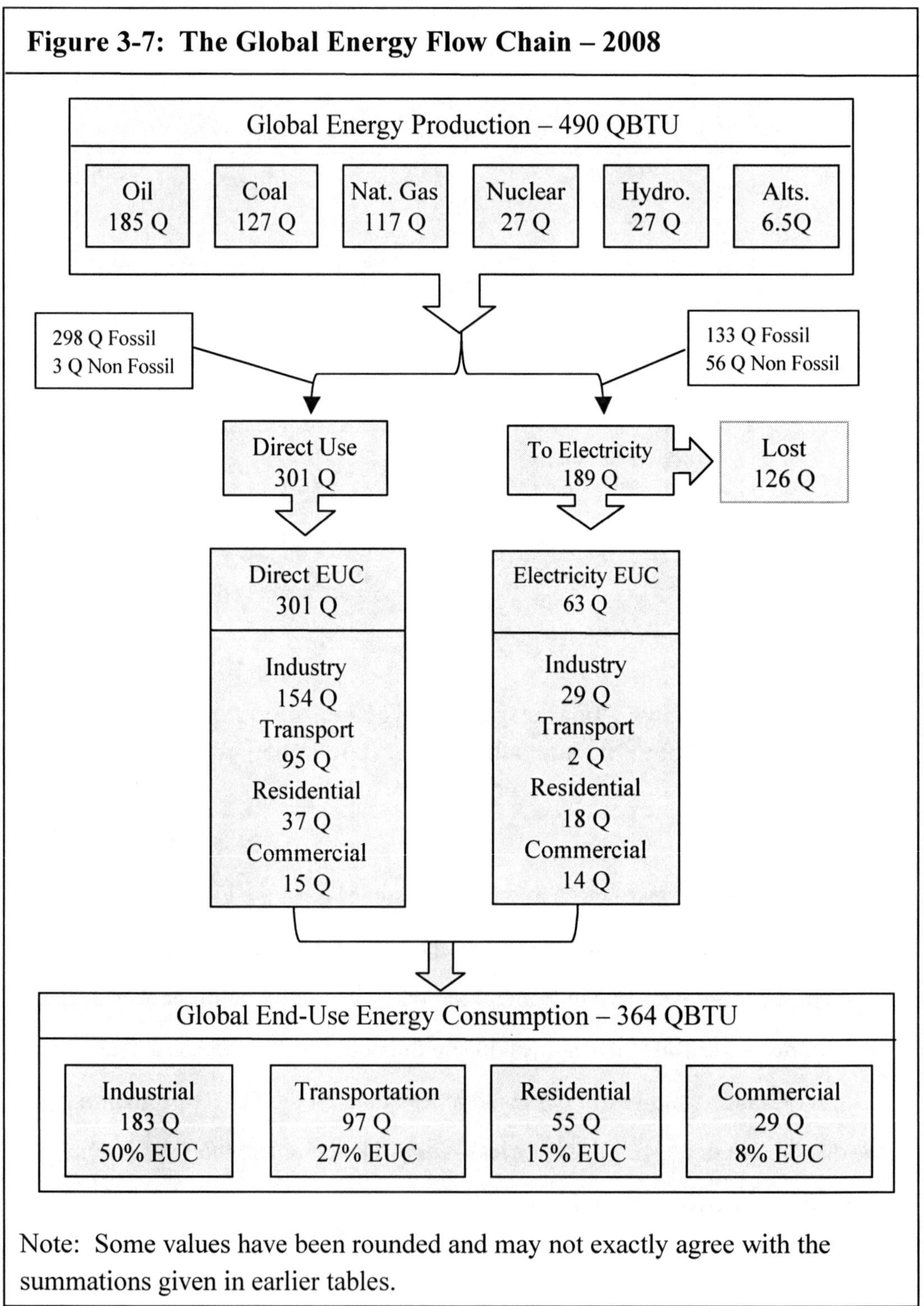


Note: Some values have been rounded and may not exactly agree with the summations given in earlier tables.

## Section 3: Where Does All the Energy Go?

### The Energy Flow Chain – the Big Picture

Figure 3-7 gives the final level of detail for the energy-flow chain. The figure tracks the primary energy sources through the direct-use and electricity generation paths and on to the end-use consumption sectors. From the flow chain we can see that direct-energy use (301 QBTU) is almost all fossil and the electricity path is about 70 percent fossil fuel (133 out of 189 QBTU).

*The Most Important Figure*

Figure 3-7 is the most important figure in this handbook in terms of usefulness. With it you will be able to verify or challenge almost any energy-related statement that comes your way. In Section 9 several examples are given which use the energy chain to interpret several of the many of the energy-saving proposals that you have heard about.

*Using the Flow Chain*

The following shows how the energy-flow chain can be used to give a quick assessment of a particular element of global energy consumption. From the energy chain, residential electricity consumption is seen to be 18 QBTU. This represents about 5 percent of global EUC (18 out of 364 QBTU). The equivalent of this in PSE is about 54 QBTU (three times the EUC value) or about 11 percent of all global energy (54 out of 490 QBTU). This points out the problem of trying to conserve energy at the global level. Heroic efforts by homeowners to save electricity will have a relatively small effect on global energy consumption. If every household in the world could reduce its electricity use by 15 percent, the result would be only a 1.7 percent decrease in total global energy use (15% of 54 divided by 490 QBTU). With electricity consumption increasing at about 2.5 percent yearly, even this small gain would be short-lived.

## Section 3: Where Does All the Energy Go?

### How Much Goes Into?

Now that you have a feeling of where energy comes from and where it goes, here is a game that you can play with your friends when things get really slow. It's called "How Much Gozinto." Hint: you can use the energy-flow chain, but don't give a copy to your friends – let them guess.

*World Total Energy Production*

Q. What percent of world energy (PSE) goes into electricity generation?
A. 39 percent – 189 QBTU out of 490 QBTU.

Q. What percent of world PSE reaches electricity end-use consumers?
A. 13 percent – 63 QBTU out of 490 QBTU.

Q. What percent of world PSE goes into the transportation sector?
A. 20 percent – 97 QBTU out of 490 QBTU.

Q. What percent of world PSE goes into the motor vehicle sub-sector?
A. 12 percent – 60 QBTU out of 490 QBTU.

*World Fossil-Fuel Production*

Q. What percent of world fossil fuel goes into electricity generation?
A. 31 percent – 133 QBTU out of 429 QBTU.

Q. What percent of world fossil fuel goes into the transportation sector?
A. 23 percent – 97 QBTU out of 429 QBTU.

Q. What percent of world fossil fuel goes into the motor vehicle sub-sector?
A. 14 percent – 60 QBTU out of 429 QBTU.

## Section 3: Where Does All the Energy Go?

*Electricity Generation*

Q. What percent of fossil fuel goes into the electricity generation mix?
A. 70 percent – 133 QBTU out of 189 QBTU.

Q. What percent of world PSE to electricity reaches EUC?
A. 33 percent – 63 QBTU out of 189 QBTU.

Q. What percent of world PSE to electricity is lost before reaching EUC?
A. 67 percent – 126 QBTU out of 189 QBTU.

Well, maybe this is not such a good game to play with your friends after all. The real point here is to show how confusing the energy percentage game can be. There are endless combinations of a percentage of something compared to something else. I am sure you have run across energy reports that go on for pages giving percentages of this to that, but with no frame of reference that might help you interpret the information. Keep your copy of the energy flow chain handy. With it you will be able to quickly assess the meaning of most energy percentages that you will ever encounter. If not, at least you will be able to ask questions like "What is this a percentage of ?" or "Is this a percentage of total PSE or EUC?" or "Is this a percentage of fossil fuel?" or "Is this a percentage of a major sector?" or " Is this a percentage of something within a particular sub-sector?"

**End-Use Consumption Growth Rate**

Figure 3-8 shows total world EUC projections to the year 2030. The figure shows that global energy consumption is increasing by about two percent yearly and is predicted to continue at this rate through the year 2030. Table 3-3 gives the data presented in Figure 3-8 in tabular form.

## Section 3: Where Does All the Energy Go?

Figure 3-8 also shows the projected EUC growth rates for the four major consumption sectors. No surprises here; each sector is predicted to have a consistent growth rate to the year 2030. The transportation sector, which is almost completely dependent on petroleum, does not show any tailing-off of its growth rate despite the threat of diminished supplies during the years leading up to 2030. It should be noted that drastic changes in the global oil supply due to geopolitical, nationalistic or global warming-driven treaty restrictions could dramatically alter these EIA predictions which are based a nominal "two-percent" growth rate.

Figure 3-8: World Energy End-Use Consumption by Sector (from EIA Figure 2)

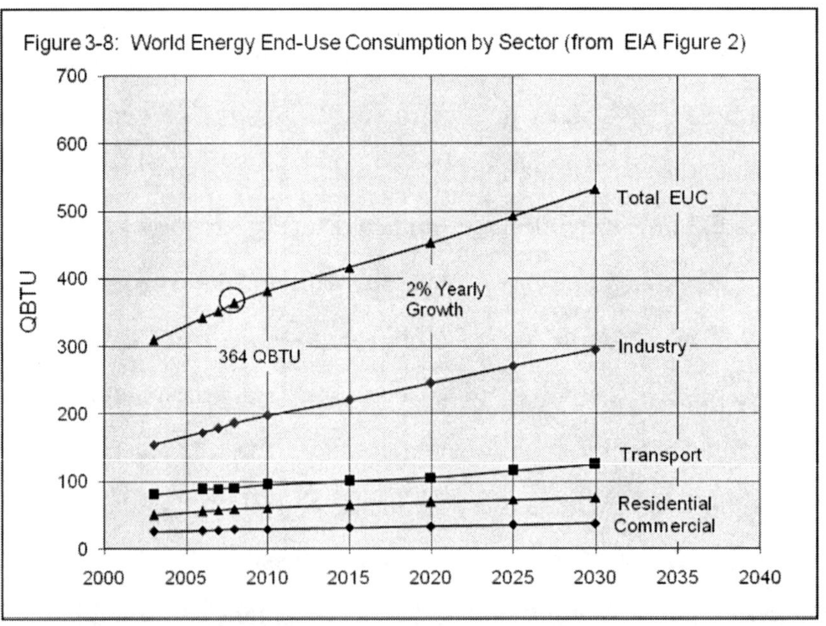


| Table 3-3: World End-Use Consumption by Year (from EIA Figure 2) | | | | | |
|---|---|---|---|---|---|
| Year | Industry | Transport | Residential | Commer. | Total |
| 2003 | 154 | 80 | 50 | 25 | 309 |
| 2006 | 172 | 88 | 55 | 26.5 | 342 |
| 2007 | 178 | 89 | 57 | 27.3 | 351 |
| **2008** | **186.4** | **90.4** | **58.4** | **28.4** | **364** |
| 2010 | 197 | 95 | 60 | 29 | 381 |
| 2015 | 220 | 100 | 65 | 31 | 416 |
| 2020 | 245 | 105 | 69 | 33 | 452 |
| 2025 | 270 | 115 | 72 | 35 | 492 |
| 2030 | 295 | 125 | 75 | 37 | 532 |

## Section 3: Where Does All the Energy Go?

### Prime Source Energy by the Sectors

Figure 3-4 showed the distribution of EUC between the major sectors. In Figure 3-9 we now see how PSE, which includes energy losses, is distributed between the sectors. The darkly shaded areas at the bottom of each bar represent the approximate amount of energy lost in each sector. The losses result mainly from fossil-fuel energy conversion to electricity plus subsequent transmission losses to end-use consumers. Together the losses add up to126 QBTU as seen in the energy-flow chain (Figure 3-7). The small energy loss shown for the transportation sector is reflective of the very small amount of electricity used by this sector.

Figure 3-9: Prime Source Energy Use by the Sectors

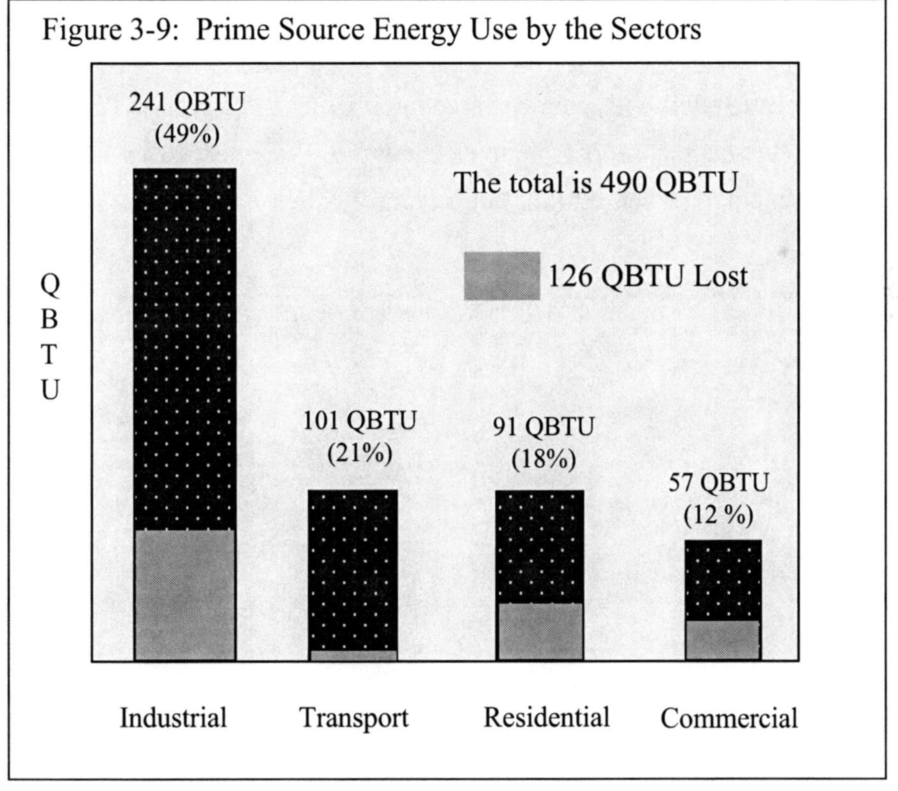

## Section 3: Where Does All the Energy Go?

### Summary of Section 3


The *Global Energy Flow Chain* (Figure 3-7) is your most important tool for interpreting global energy information.

The *End-Use Consumption* chart (Table 3-2) shows where energy is consumed down to the sector and sub-sector levels.

Nearly 40 percent of global prime source energy goes into the electricity generation process, but only about a third of this reaches the end-use consumer.

End-use energy consumption is dominated by the industrial, transportation and commercial sectors, while we at home (the residential sector) account for only 15 percent of consumed energy.

When confronted with energy percentage values, always ask "Percentage of what?" It is nearly impossible to read anything on energy that does not give it as a percentage of something, but it is rarely clear as to the percentage of what.

# Section 4
# Energy in Everything

As pointed out in the beginning of this handbook, all sources and forms of energy can be expressed in common set of energy units. The BTU was chosen as the common unit of energy. Figure 4-1 gives the BTU per pound (specific energy) for each of the prime energy sources covered in Section 2. Electrical energy itself cannot be given in terms of specific energy, but the energy stored in a battery can be expressed on a per-pound basis.

## Energy Content of the Prime Sources

Because the BTU is a relatively small unit, Figure 4-1 gives the specific energy values for the several energy sources in thousands of BTU or KBTU per unit weight. From the figure it is seen that hydrogen gas contains the highest BTU per pound of all the energy sources. Hydrogen, however, is not a prime source of energy; it must be produced from some other source, like natural gas or coal. It is therefore referred to as an energy-carrier. The specific energy content of the modern lithium-ion battery is included with the other energy sources for the purpose of giving a relative energy storage comparison. Yes, something is there for the lithium-ion battery, it's just a little difficult to see.

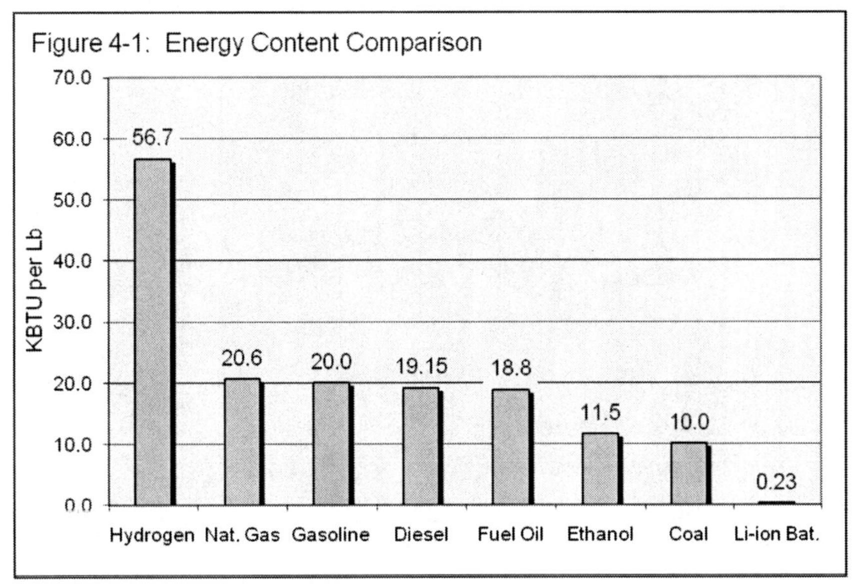

## Section 4: Energy in Everything

Natural gas and petroleum derivatives have approximately the same energy content, about 20 KBTU per pound. Coal, as common and useful as it is, has only about half the energy content of petroleum or natural gas, about 10 KBTU per pound. The energy content of ethanol and the other biofuels, at about 12 KBTU per pound, come in slightly ahead of coal, but considerably behind petroleum and its derivatives. The figure points out the tremendous energy-storing capabilities of the fossil fuels compared to even the most advanced lithium-ion battery. The energy capacity of the lithium-ion battery, 0.23 KBTU per pound, is around 100 times less than that of natural gas or the petroleum derivatives.

**Energy Reference Tables**

The tables in this section provide a quick and convenient database for each of the world's prime energy sources. The data is composite in nature, collected from multiple Internet sources and is representative of the 2007-08 time frame. Included are the more important parameters that you should find useful for making comparisons between the various prime energy sources.

- Petroleum – Mr. Big (for now)
    - Gasoline – Makes the World Go
    - Diesel Fuel – Big Rig Friend
    - Fuel Oil # 2 – Heating It Up
- Natural Gas – Low-Profile Worker
- Coal – Energy Workhorse
- Nuclear – Silent Worker
- Hydroelectric – Water over the Dam
- Biofuels – Home Grown
- Geothermal – Deep Heat
- Wind – Big Breeze
- Solar – A Ray of Hope?

## Section 4: Energy in Everything

The tables include basic data for each of the primary energy sources. Data is first given in familiar energy working units: barrels of oil, gallons of gasoline, cubic feet of natural gas, short tons of coal, etc. The tables also give energy per unit weight, cost per unit energy and total world energy production in percentage terms. The working units are then converted to BTU to provide for a common denominator reference between each of the prime sources. If nothing else these tables will save you hours of downloading and interpreting the often ambiguous or incomplete energy information found on the Internet.

### Petroleum: Mr. Big (for now)

Petroleum is really the Mr. Big of the energy world. Petroleum provides 38 percent of the global energy supply. The energy content of oil is about 19 KBTU per pound in its crude state. The uses for petroleum are obvious and too numerous to list. Most oil still comes in the black liquid form that we have become familiar with, however an increasing percentage is beginning to come from bituminous sands (aka tar or oil sand), mainly from large resources located in Canada and Venezuela (the good news). The energy required (mainly from natural gas) to process bituminous sand into a useful liquid form is several times greater than for the refinement of traditional crude oil (the bad news).

| Table 4-1: Petroleum | Value | Units |
|---|---|---|
| Energy per barrel | 5.86E+06 | BTU/bbl |
| Gallons oil per barrel (U.S. gallon) | 42 | gal. oil/bbl |
| Energy per gallon | 1.39E+05 | BTU/gal |
| Energy per unit weight | 1.88E+04 | BTU/lb |
| Density | 55.6 | lb/ft3 |
| World barrels per day | 8.64E+07 | bbl/day |
| World barrels per year | 3.15E+10 | bbl/yr. |
| World gallons per day | 3.63E+09 | gal. oil/day |
| World gallons per year | 1.32E+12 | gal. oil/yr. |
| World energy, BTU per year | **1.85E+17** | BTU/yr. |
| Above in percent of world energy | 37.7% | % |
| Dollars per barrel (mid 2008) | $125 | $/bbl |
| Dollars per million BTU | $21.35 | $/MBTU |

## Section 4: Energy in Everything

The world used over 31 billion barrels of oil in 2007. Figure 4-1a show what a barrel of oil yields after going through the refining process.

Figure 4-1a: Where a Barrel of Oil Goes

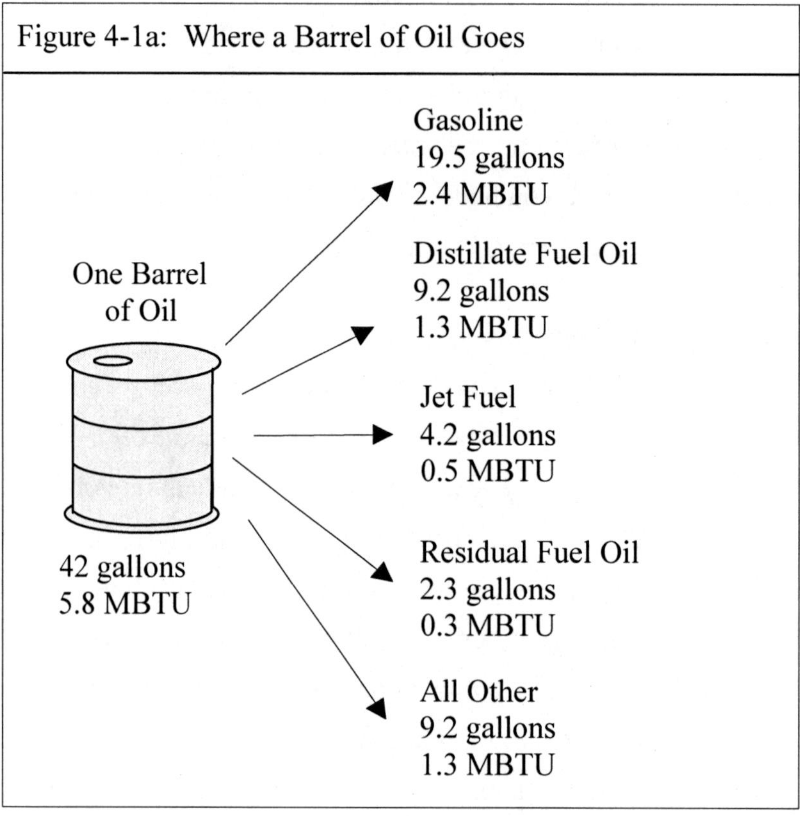


It would be convenient if each barrel of oil could be converted to any amount of desired end-product, gasoline or aviation fuel for example, but it doesn't work that way. The distillation-refining process yields a range of petroleum end-products – from high-quality gasoline to lower grade fuel oil, to source materials used in the petrochemical industry – in about the amounts as shown in the figure.

Figure 4-1b shows the growth of petroleum consumption through 2030 to be in the range of 1.5 percent yearly – shortages, rising costs, world events and environmental concerns notwithstanding.

## Section 4: Energy in Everything

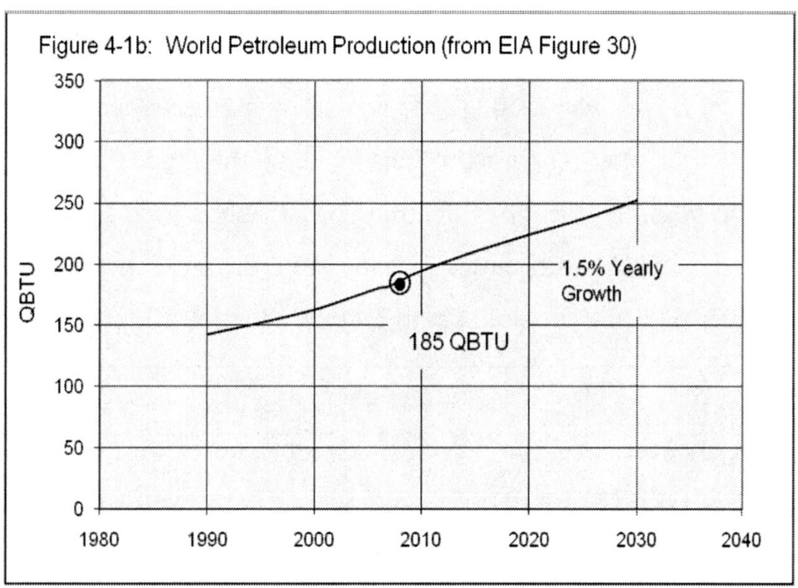


Figure 4-1c shows the transportation and industrial sectors leading the way in projected petroleum consumption. The residential and commercial sectors are expected to remain relatively small users of petroleum.

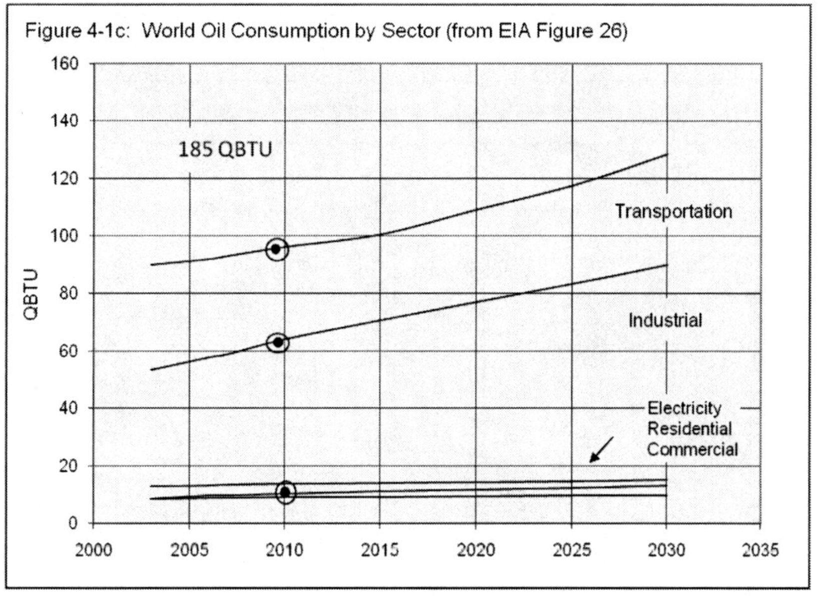

## Section 4: Energy in Everything

### *Petroleum Derivative – Gasoline*

Over forty percent of the world's oil production ends up as gasoline (75 out of 185 QBTU). Of this about 60 QBTU is used by motor vehicles. Gasoline along with aviation, diesel and other fuels make up the total of 95 QBTU used by the transportation sector. The remaining petroleum is used for residential heating and industrial-petrochemical purposes. Figure 4-1d shows gasoline and non-gasoline as part of total global petroleum production.

| Table 4-1a: Gasoline | Value | Units |
|---|---|---|
| Gallons gasoline per barrel of oil | 19.5 | gal. gas/bbl oil |
| Energy per gallon (U.S. gallons) | 1.25E+05 | BTU/gal |
| Density | 46 | lb/ ft3 |
| Energy per unit weight | 2.03E+04 | BTU/lb |
| Energy per unit volume | 9.35E+05 | BTU/ft3 |
| Gasoline barrels per day | 3.79E+07 | bbl/day |
| Gasoline barrels per year | 1.38E+10 | bbl/yr. |
| Gasoline gallons per day | 1.64E+09 | gal./day |
| Gasoline gallons per year | 6.00E+11 | gal./yr. |
| World energy, BTU per year | **7.5E+16** | BTU/yr. |
| Above in percent of world energy | 15.3% | % |
| Dollars per gallon (early 2008) | $4.25 | $/gal. |
| Dollars per million BTU | $34.00 | $/MBTU |

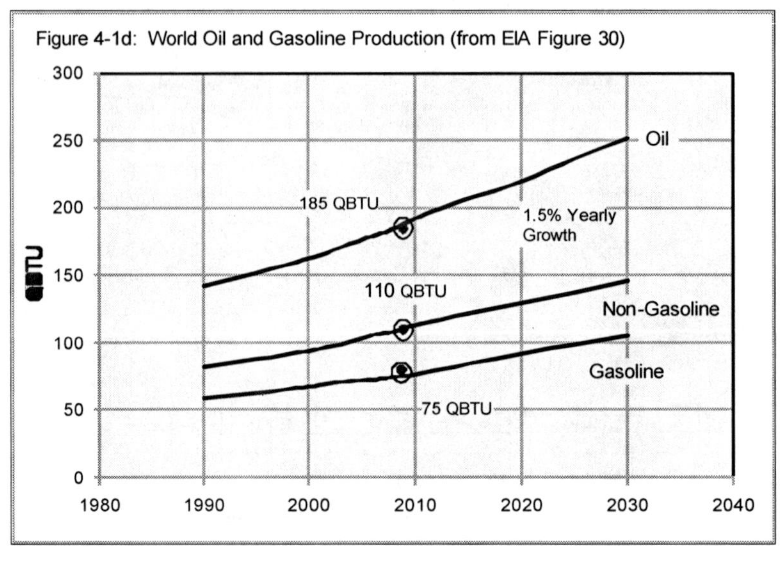

## Section 4: Energy in Everything

Tables 4-1d and 4-1e give year-by-year data on petroleum and gasoline production in the energy units commonly used in the media and also in QBTU. This table should help you when reading articles that may give oil and gasoline data in any of the forms listed. There seems to be no consistent way of reporting this in the media. Often you will encounter two or more of these unit systems being used within the same article or report. Getting the units straight can be just as important as getting the correct energy values.

**Table 4-1d: World Oil Production**

| Year | bbl./day | bbl./yr. | gal./day | gal./yr. | QBTU/yr. |
|---|---|---|---|---|---|
| 1990 | 6.65E+07 | 2.43E+10 | 2.79E+09 | 1.02E+12 | 142 |
| 1995 | 7.10E+07 | 2.59E+10 | 2.98E+09 | 1.09E+12 | 152 |
| 2000 | 7.61E+07 | 2.78E+10 | 3.20E+09 | 1.17E+12 | 163 |
| 2003 | 8.00E+07 | 2.92E+10 | 3.36E+09 | 1.23E+12 | 171 |
| 2006 | 8.40E+07 | 3.07E+10 | 3.53E+09 | 1.29E+12 | 180 |
| 2007 | 8.50E+07 | 3.10E+10 | 3.57E+09 | 1.30E+12 | 182 |
| **2008** | **8.65E+07** | **3.16E+10** | **3.63E+09** | **1.33E+12** | **185** |
| 2010 | 9.00E+07 | 3.29E+10 | 3.78E+09 | 1.38E+12 | 192 |
| 2015 | 9.70E+07 | 3.54E+10 | 4.07E+09 | 1.49E+12 | 207 |
| 2020 | 1.03E+08 | 3.76E+10 | 4.33E+09 | 1.58E+12 | 220 |
| 2025 | 1.11E+08 | 4.05E+10 | 4.66E+09 | 1.70E+12 | 237 |
| 2030 | 1.18E+08 | 4.31E+10 | 4.96E+09 | 1.81E+12 | 252 |

**Table 4-1e: World Gasoline Production**

| Year | bbl./day | bbl./yr. | gal./day | gal./yr. | QBTU/yr. |
|---|---|---|---|---|---|
| 1990 | 2.93E+07 | 1.07E+10 | 1.30E+09 | 4.76E+11 | 59 |
| 1995 | 3.12E+07 | 1.14E+10 | 1.39E+09 | 5.08E+11 | 63 |
| 2000 | 3.35E+07 | 1.22E+10 | 1.49E+09 | 5.44E+11 | 68 |
| 2003 | 3.52E+07 | 1.28E+10 | 1.57E+09 | 5.72E+11 | 72 |
| 2006 | 3.70E+07 | 1.35E+10 | 1.65E+09 | 6.01E+11 | 73 |
| 2007 | 3.74E+07 | 1.37E+10 | 1.65E+09 | 6.02E+11 | 74 |
| **2008** | **3.81E+07** | **1.39E+10** | **1.70E+09** | **6.19E+11** | **75** |
| 2010 | 3.96E+07 | 1.45E+10 | 1.76E+09 | 6.44E+11 | 77 |
| 2015 | 4.27E+07 | 1.56E+10 | 1.90E+09 | 6.94E+11 | 85 |
| 2020 | 4.53E+07 | 1.65E+10 | 2.02E+09 | 7.37E+11 | 92 |
| 2025 | 4.88E+07 | 1.78E+10 | 2.18E+09 | 7.94E+11 | 99 |
| 2030 | 5.19E+07 | 1.90E+10 | 2.31E+09 | 8.44E+11 | 106 |

# Section 4: Energy in Everything

## Coal: The Energy Workhorse

Coal provides about 26 percent of the total global energy supply. Although it is the workhorse of industry and electricity generation, coal has a relatively low energy content. There are numerous types of coal found throughout the world, each having somewhat different energy contents. Table 4-2 gives the energy content as 10 KBTU per pound – an average for several common coal types.

Approximately two-thirds of the world's coal production is used to generate electricity. Coal makes up 40 percent of the total energy mix going into electricity. Figure 4-2a shows the past history and future coal production predicted through 2030.

| Table 4-2: Coal | Value | Units |
|---|---|---|
| Energy per short ton (averaged types) | 2.00E+07 | BTU/short ton |
| Energy per unit weight | 10,000 | BTU/lb avg. |
| World short tons per year | 6.35E+09 | short tons/yr. |
| World pounds per year | 1.27E+13 | lb |
| World energy, BTU per year | **1.270E+17** | BTU/yr. |
| Above in percent of world energy | 25.9% | % |
| Percent coal to electricity generation | 66% | % |
| Percent coal in electricity input mix | 40% | % |
| Dollars per short ton (2008 avg., elec. power) | $35 | $/short ton |
| Dollars per million BTU | $1.75 | $/MBTU |

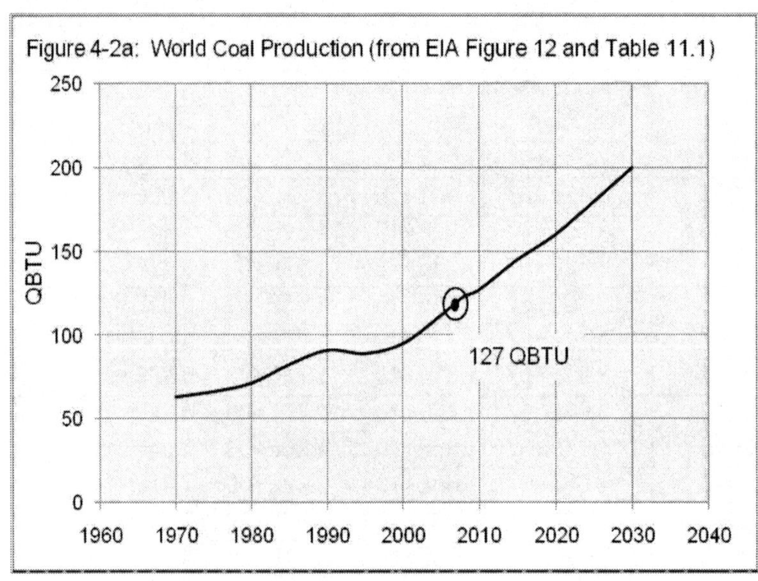

## Section 4: Energy in Everything

*Coal Energy Data Spread*

Figure 4-2b has been included here to point out the data spread that one may encounter when searching the Internet for energy information. Quite often data spreads are due to the definition of terms. In this case, note the labels on the curves: “World Marketed Energy,” “Coal Consumption,” “World Coal Production.” Each label probably has a slightly different meaning to those in the coal industry which could explain the data spread. A nominal value of 127 QBTU per year was selected for this work as the reference value for coal production during the 2007-08 period.

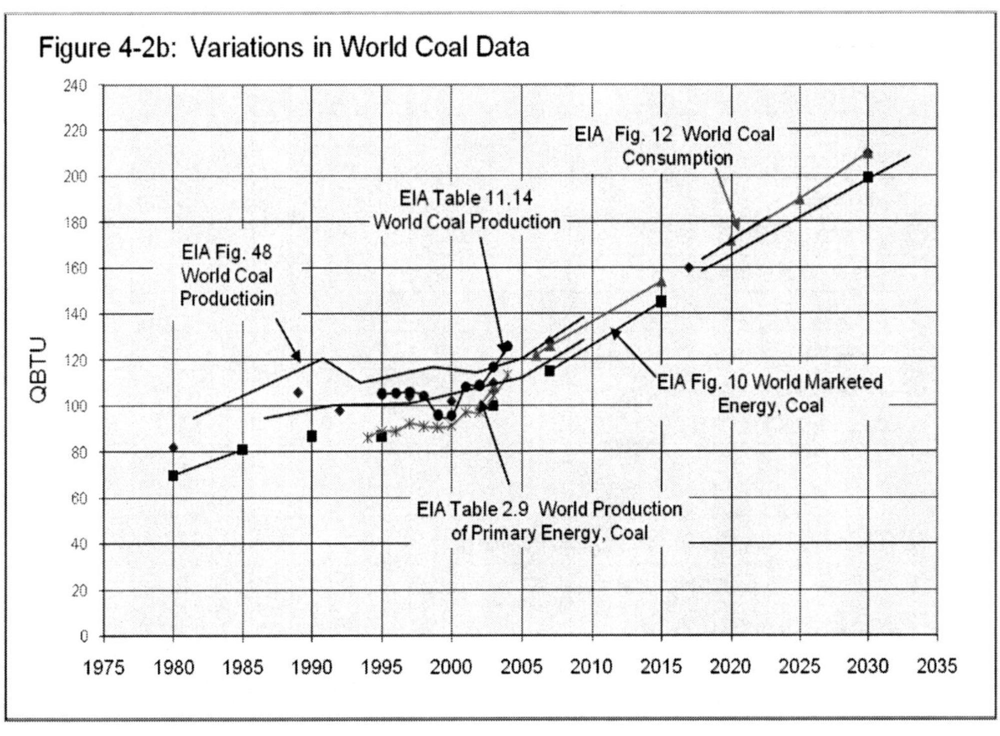

## Section 4: Energy in Everything

### Natural Gas: Low-Profile Worker

Natural gas provides nearly 24 percent of the world's prime energy (Table 4-3). Its primary uses are for heating, both residential and commercial, and for conversion to electricity – a third of all natural gas goes into electricity production. Natural gas represents 20 percent of the energy mix going into global electricity generation.

The numerous ways in which natural gas energy is measured can be confusing. Table 4-3a is included to help make things a little easier when trying to convert mixed-unit natural gas data into BTU.

| Table 4-3: Natural Gas | Value | Units |
|---|---|---|
| Energy per therm | 1.00E+05 | BTU/therm |
| Density | 0.05 | lb/ft3 |
| Therms per cubic foot | 1.03E-02 | therm/ft3 |
| Energy per unit weight | 2.06E+04 | BTU/lb |
| Energy per unit volume | 1030 | BTU/ft3 |
| World energy, cubic ft | 1.14E+14 | ft3/yr. |
| World energy, BTU per year | **1.174E+17** | BTU/yr. |
| Above in percent of world energy | 23.9% | % |
| Percent nat. gas to electricity generation | 33% | % |
| Percent nat. gas in electricity input mix | 20% | % |
| Dollars per therm (early 2008) | $1.50 | $/therm |
| Dollars per million BTU | $15.00 | $/MBTU |

| Table 4-3a: Natural Gas Conversions | | |
|---|---|---|
| Multiply This | By | To Get |
| Cubic Feet Natural Gas | 1030 | BTU |
| BTU | 0.0010 | Cubic Feet |
| Therms | 1.00E+05 | BTU |
| BTU | 1.00E-05 | Therms |
| Cubic Feet | 1.030E-02 | Therms |
| Therms | 97 | Cubic Feet |

## Section 4: Energy in Everything

The demand for natural gas is predicted to continue at a steady pace through 2030 as indicated in Figure 4-3.

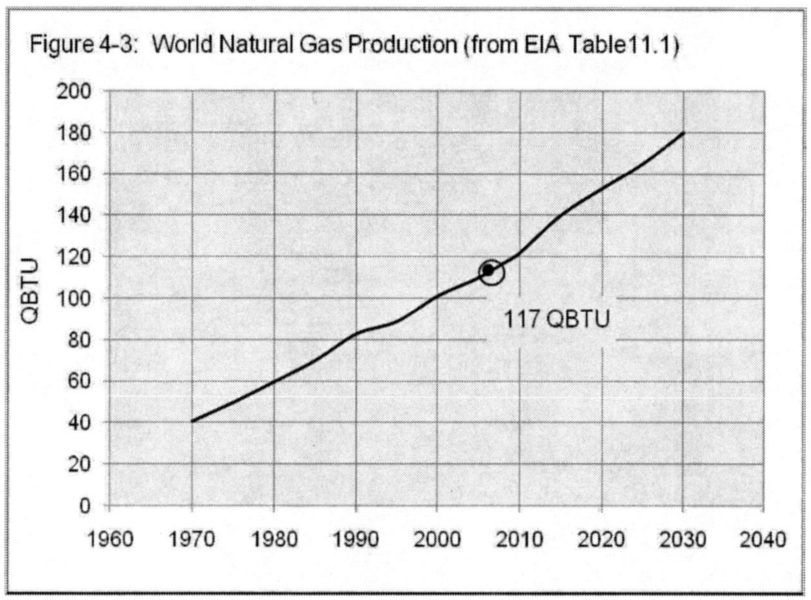


### Nuclear Energy: The Silent Worker

Nuclear energy accounts for only 6 percent of the world's PSE, but supplies 15 percent of the energy input to global electricity generation (Table 4-4). Nuclear has the potential to replace a significant portion of the 133 QBTU of fossil fuel (coal and natural gas) now used to generate electricity. However it would be of little value for reducing our dependence on petroleum – contrary to what is often heard in the media and by some politicians.

| Table 4-4: Nuclear Energy | Value | Units |
|---|---|---|
| World power, megawatts | 360,600 | MW |
| Percent nuclear to electricity | 100% | % |
| Percent nuclear in electricity input mix | 15% | % |
| World energy, megawatt-hour per year | 8.00E+09 | MWh/yr. |
| World energy, BTU per year | **2.7.E+16** | BTU/yr. |
| Above in percent of world energy | 5.6% | % |

## Section 4: Energy in Everything

Nuclear energy is predicted to be on a steady, if not spectacular, rise even with the issues of safety and waste disposal being unresolved in the public mind (Figure 4-4).

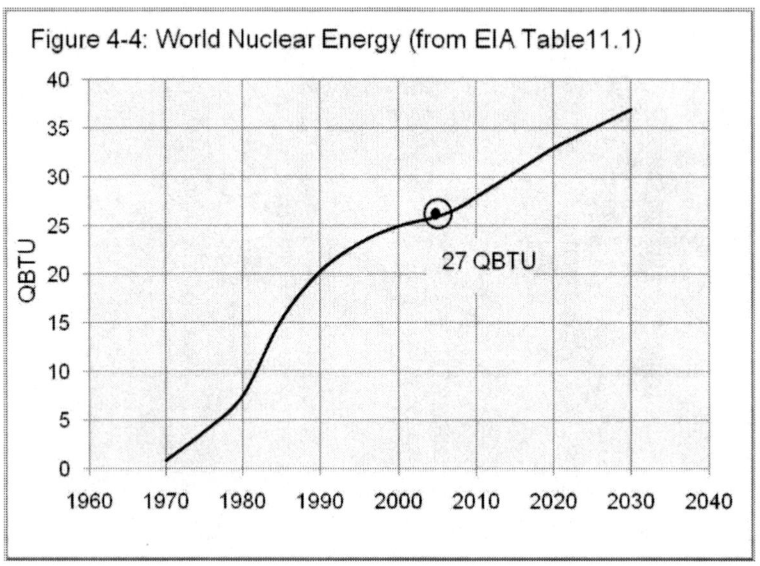


### Hydroelectric Energy: Water Over the Dam

Hydroelectric power plants, like nuclear power plants, supply about 6 percent of the world's prime source energy, but provide 15 percent of the input energy to global electricity generation (Table 4-5).

| Table 4-5: Hydroelectric Energy | Value | Units |
|---|---|---|
| World power, megawatts | 900,000 | MW |
| Percent hydroelectric to electricity | 100% | % |
| Percent hydroelectric in electricity input mix | 15% | % |
| World energy, megawatt-hour per year | 8.00E+09 | MWh/yr |
| World energy, BTU per year | **2.7E+16** | BTU/yr |
| Above in percent of world energy | 5.6% | % |

Hydroelectric energy is predicted to be on a steady rise (Figure 4-5) even though new sites are geographically limited and are often opposed on environmental grounds.

## Section 4: Energy in Everything

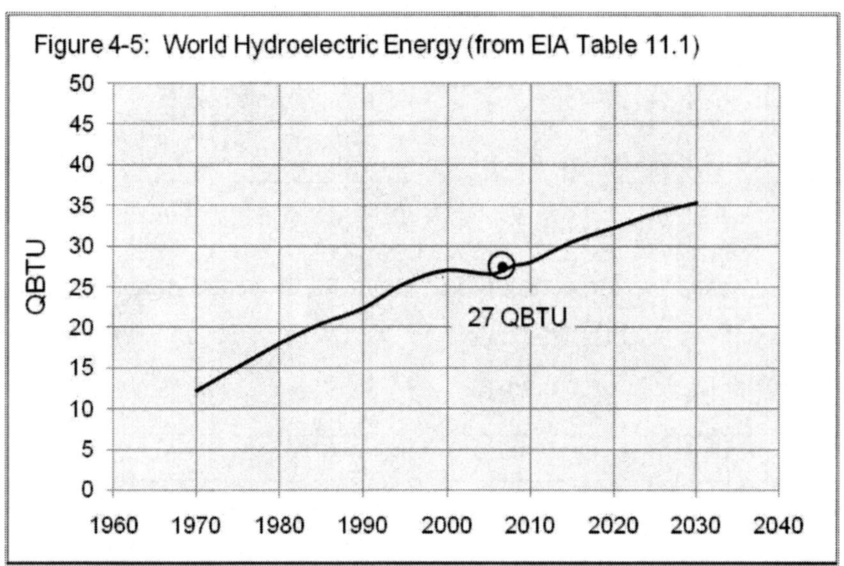


### Biofuels: Home Grown

Biofuels (ethanol and biodiesel) currently make up less than 0.25 percent of world prime source energy. Ethanol production (Table 4-6a) is estimated to be about 12 billion gallons or about 1 QBTU for 2008. The country of Brazil itself produced about half of the world's ethanol in 2007. The family of biomass fuels – wood, brush and agricultural waste products – is roughly estimated to be in the range of 3 to 4 QBTU or about one percent of global energy (Table 4-6b). Obtaining a consistent set of data on biomass fuel is difficult because of their many types, and also because worldwide reporting and documentation is not as detailed as for petroleum or the other fossil fuels.

| Table 4-6a: Ethanol | Value | Units |
|---|---|---|
| Energy per gallon | 7.60E+04 | BTU/gal |
| Density | 49.2 | lb/ft3 |
| Energy per unit weight | 1.15E+04 | BTU/lb |
| World production, gallons per year | 1.20E+10 | gal/yr. |
| World energy, BTU per year | **9.12E+14** | BTU/yr. |
| Above in percent of world energy | 0.19% | % |
| Dollars per gallon (early 2008, very approx.) | $3.75 | $/gal |
| Dollars per million BTU | $49.34 | $/MBTU |

## Section 4: Energy in Everything

| Table 4-6b: Biomass | Value | Units |
|---|---|---|
| World energy, BTU per year | **4.30E+15** | BTU/yr. |
| Above in percent of world energy | 0.88% | % |

### Wind Energy: The Big Breeze

Wind-generated power, estimated at about 90 thousand megawatts globally for 2007, provides slightly less than one QBTU per year, or about a 0.2 percent of the world's prime source energy.

| Table 4-7: Wind Energy | Value | Units |
|---|---|---|
| World power, megawatts | 90,000 | MW |
| Operational hours per year | 3022 | hr/yr. |
| Percent operational time per year | 35% | % |
| World energy, megawatt-hour per year | 2.72E+08 | MWh/yr. |
| World energy, BTU per year | **9.28E+14** | BTU/yr. |
| Above in percent of world energy | 0.19% | % |

### Geothermal Energy: Deep Heat

Geothermal power is estimated at 40,000 megawatts – combined electrical and thermal equivalents. Currently geothermal sources supply only about 0.8 QBTU or 0.17 percent of the world's energy. This is expected to markedly increase with improvements in extraction technology. The estimate for future *extractable* geothermal energy is astronomical – from 200 to 2000 ZJ! No need to convert this to anything else, it's thousands of times more energy than we will ever need, although this may be a bit far down the time tunnel (Ref 13, IEA).

| Table 4-8: Geothermal Energy | Value | Units |
|---|---|---|
| World power, megawatts | 40,000 | MW |
| Operational hours per year | 6132 | hr/yr. |
| Percent operational time per year | 70% | % |
| World energy, megawatt-hour per year | 2.45E+08 | MWh/yr. |
| World energy, BTU per year | **8.37E+14** | BTU/yr. |
| Above in percent of world energy | 0.17% | % |

## Section 4: Energy in Everything

This 200 to 2000 ZJ statement for geothermal energy is similar to the often heard statement about solar energy – "Enough sunlight falls on the earth to supply all of our energy needs forever."

### Solar Energy: Ray of Hope?

Solar power is probably the most talked about and controversial alternative energy source. To many it is the answer to the world's energy problems, to others it is only a partial solution at best. Currently solar energy provides a very small portion (0.01 percent) of world's energy supply as indicated in the Table 4-9. Solar power capability for 2007-08 is estimated to be in the range of 9,000 megawatts. This roughly translates to 20 million megawatt-hours or about 0.07 QBTU.

Because solar systems are located in wide range of geographic regions, each with varying hours-per-year of sunlight, there is no one exact factor for converting from megawatts of solar power to QBTU of solar energy. For Table 4-9 an operational effectiveness factor of 25 percent is used. This is based on 365 days, 8 hours per day of sunlight and a 0.75 clear weather factor. The eight hours-per-day sunlight is an average for summer and winter. Yes, there are many regions in the world where the operational effectiveness factor is higher, but there also many regions where it is much lower. There are numerous Websites providing information on solar energy effectiveness for your region of the country or world.

| **Table 4-9: Solar Energy** | **Value** | **Units** |
|---|---|---|
| World power, megawatts | 9,000 | MW |
| Operational hours per year | 2190 | hr/yr. |
| Percent operational time per year | 25% | % |
| World energy, megawatt-hour per year | 1.97E+07 | MWh/yr. |
| World energy, BTU per year | **6.73E+13** | BTU/yr. |
| Above in percent of world energy | 0.01% | % |

# Section 5
## Energy on the Home Front

In this section we will be looking at energy around the home, which of course means natural gas, fuel oil and electricity. An interesting way to look at household energy costs is to compare them with the costs of the other forms of energy that we use.

### The Cost of Energy

We are all familiar with thinking dollars per gallon for gasoline, but here we are looking at dollars per BTU instead. Figure 5-1 shows how many dollars it takes to buy a million BTU (MBTU) of prime energy. The figure is based on typical 2007-08 consumer energy costs as listed in Figure 5-1a.

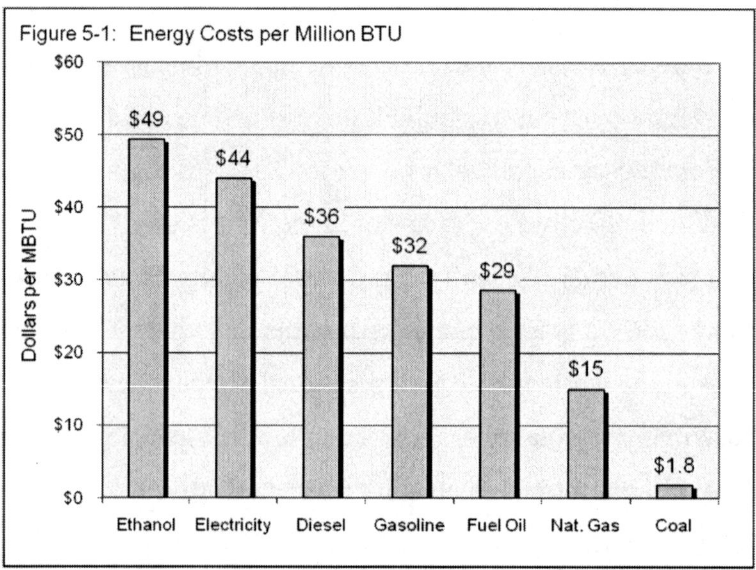


| Figure 5-1a: Consumer Energy Costs | | | |
|---|---|---|---|
| Type | Unit Cost | Units | Dollar per MBTU |
| Ethanol | $3.75 | Gallon | $49 |
| Electricity | $0.15 | kWh | $44 |
| Diesel | $5.00 | Gallon | $36 |
| Gasoline | $4.00 | Gallon | $32 |
| Fuel Oil | $4.00 | Gallon | $29 |
| Nat. Gas | $1.50 | Therms | $15 |
| Coal | $35.00 | Short tons | $1.8 |

## Section 5: Energy on the Home Front

*BTU for Your Buck*

Electricity is the most expensive form of energy on a dollars-per-MBTU basis as shown Figure 5-1. This is somewhat misleading (and I will hear from the electricity lobby on this) in that electricity is a very efficient form of energy. You will need a lot less BTU to do the same job with electricity. This will become evident in the next section where the miles-per-BTU for a gasoline vehicle is compared with the miles-per-BTU for an electric vehicle.

Coal is the least-expensive form of energy from a dollar-per-MBTU standpoint. Because of this a large portion of the world's coal supply is used to produce electricity.

*Bringing It All Home*

Electricity and natural gas are the forms of energy we come closest to as residential consumers. Look around your home and count all the things that run on the energy supplied by your friendly utilities company. You will be surprised at the number. Around my house I counted over sixty items that use electricity or natural gas. My list is given in Table 5-1. You may have even more items than this.

Table 5-1: Residential Energy-Users

| | | |
|---|---|---|
| Refrigerator | Kitchen Range | Water Heater (nat. gas) |
| Freezer | Washer, clothes | Light Bulbs (55) |
| Dish Washer | Dryer, clothes | Fluorescent tubes (20) |
| Garbage Disposal | Vacuum Cleaner (2) | TVs (3) |
| Electric Oven | Electric Iron | CD Players (3) |
| Microwave Oven | Hand Tools (many) | Tape Recorders (3) |
| Toaster | Stereo System | Phonograph (2) |
| Food Processor | VCR machines (2) | Hair Dryer |
| Coffee Maker | DVD player | Phone Chargers |
| Waffle Maker | Computer | Fans (4) |
| Toaster Oven | Monitor | Timers (4) |
| Food Mixer | Printers (2) | Outdoor tools (many) |
| Furnacc (nat. gas) | Scanner | |

# Section 5: Energy on the Home Front

## The Electricity Flow Chain

Electricity, of course, is not a source of energy. It is a transient form of energy generated from the prime sources. Except for batteries, electricity must be consumed as it is generated. Figure 5-2 is a simplified diagram showing how world prime source energy is converted into electrical energy and how much of this finally reaches you, the consumer.

**Figure 5-2: World Electricity Flow Chain**

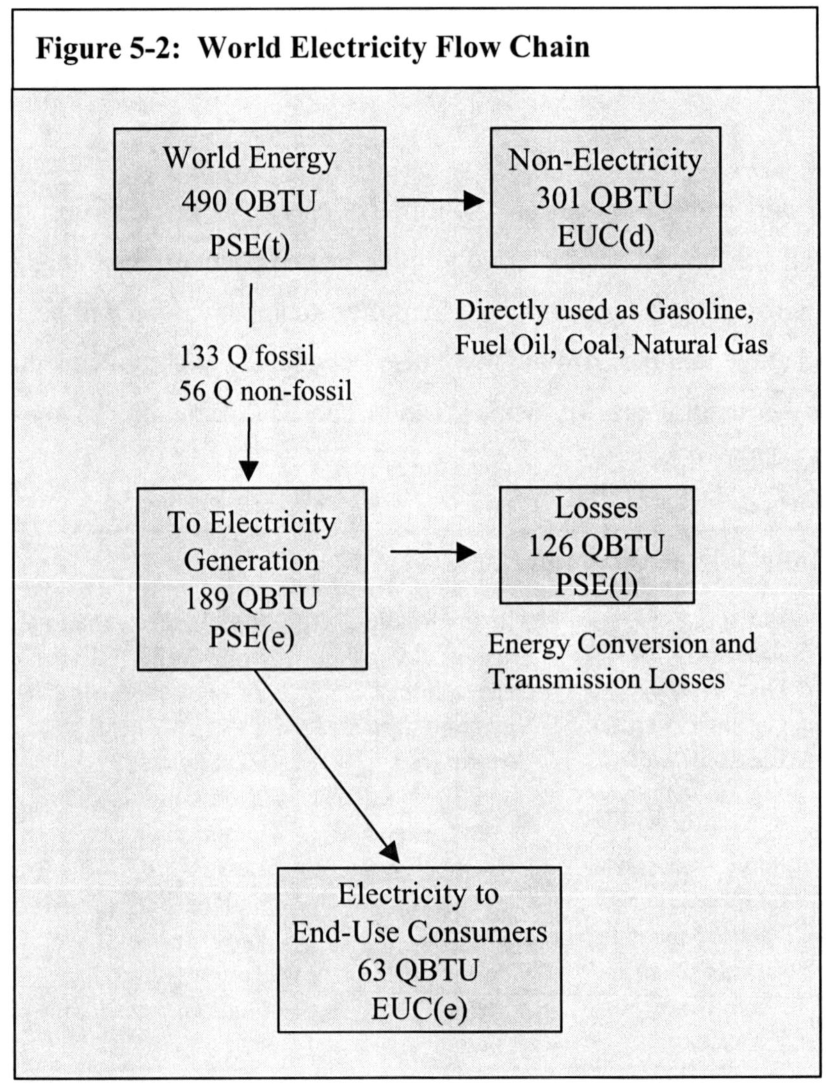

## Section 5: Energy on the Home Front

### Energy for Electricity Generation

Of the 490 QBTU of global energy produced, 189 QBTU (39 percent) is used for electricity generation. Figure 5-3 shows the past and projected growth of electricity production by type of prime-source fuel.

The energy required to generate electricity is predicted to be on a continual rise well into the foreseeable future. Coal and natural gas will fuel most of the two-percent plus yearly increase. Hydroelectric and nuclear energy will increase at lesser rates, while oil will remain a bit player in electricity generation. The alternative energy sources – solar, wind and geothermal – will be gaining on nuclear and hydroelectric energy, but will continue to be overshadowed by the fossil fuels.

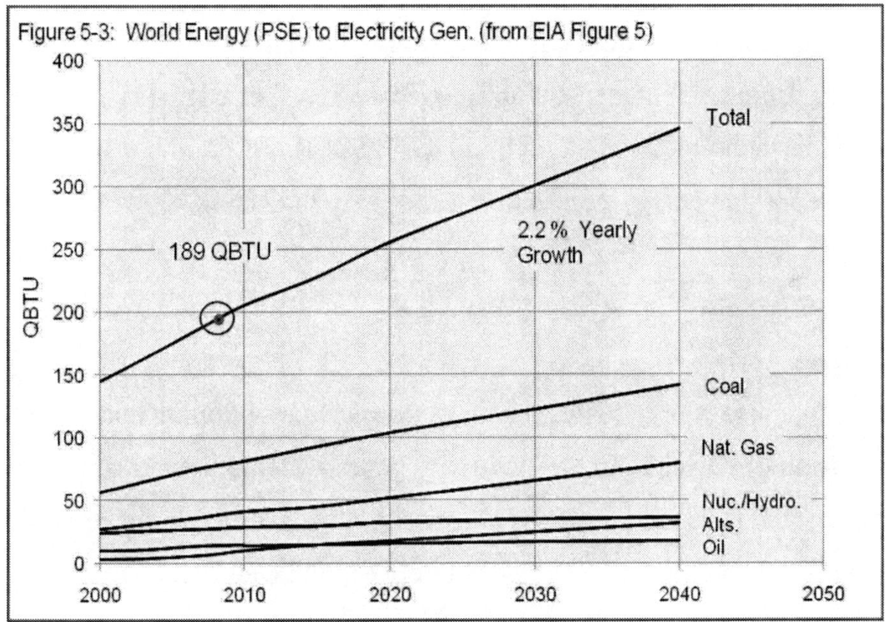


Global electricity production, fueled largely by coal and natural gas, is predicted to continue increasing at over two-percent yearly.

## Section 5: Energy on the Home Front

*Conversion Losses*

Electricity is considered one of our cleaner and cheaper forms of energy (good news). However, the conversion of coal, natural gas and oil (a small amount) into electricity is accompanied by relatively large energy losses (bad news), as was pointed out in the energy-flow chain given in Section 3. The conversion efficiencies vary depending on what country, what prime energy source and what level of technology is being used.

Efficiency factors can range between 30 to 50 percent. Coal and heating oil have conversion efficiencies at the lower end of the scale, about 33 percent. The coal gasification process has the potential to increase efficiencies above 50 percent, but because expensive facilities are required, it may be a while before this process is in general use, particularly in poorer, developing countries.

The conversion of natural gas to electricity can reach the 50-percent efficiency using combined cycle turbine technology (jet-engine type initial stage turbine combined with a traditional steam turbine). Again it will take some time for this energy-conversion technology to become the worldwide standard.

*Line Losses*

There are also energy losses resulting from the conditioning and transmission of power from the electricity generation facilities to the utilities companies and then to the end-use consumer. This is largely a function of distance and the number of relay points between the electricity generation source and your home. Most of us are not aware of these losses as we go through our daily lives. We pay our electric bill for what comes out of our wall plugs, and that's the end of it.

## Section 5: Energy on the Home Front

### *Losses Are Not Equal*

It should be noted that the electricity generation loss is an average for the total input energy mix. The mix includes not only coal and natural gas, but also hydroelectric, nuclear and alternative sources. These are direct sources of electricity and do not experience the high energy conversion losses as do the fossil fuels. The overall percentage loss for the fossils themselves (from prime source to your home receptacle) will be somewhat greater than for the average mix, maybe approaching 70 percent. Again, this will vary by country and the level of conversion technologies being used.

Electrical losses need to be taken into account whenever overall electricity efficiencies are being considered. This will become very apparent in Section 6 (Energy and Your Car), where the efficiencies of several vehicle types, including electric vehicles, are compared on a mile-per-BTU basis.

### *Electricity Pie Charts*

You have probably seen many versions of the pie chart of Figure 5-4. It shows the percentage of prime source energy by fuel type that goes into the electricity generation mix. Sometimes electricity percentages can become a little confusing. Let's look at coal, for example, to see how it can be expressed in different percentage terms.

Coal production provides *26 percent* of global prime source energy. Over *60 percent* of all coal goes toward the production of electricity. When combined with the other types of energy that go into electricity production, coal becomes *41 percent* of the total input mix. Again, whenever looking at energy percentages, especially for electricity, always ask "percentage of what?"

## Section 5: Energy on the Home Front

Figure 5-4: PSE Input to Electricity Generation

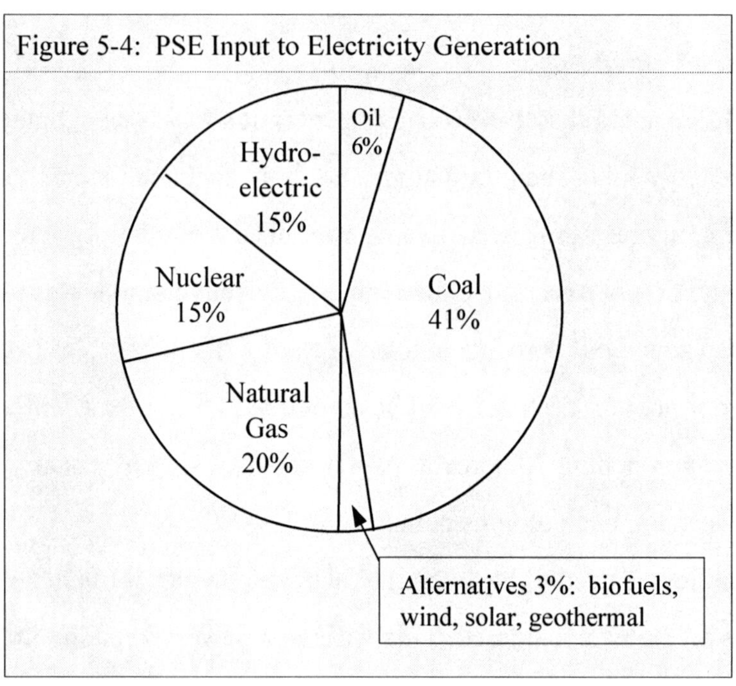


**Electricity End-Use Consumption**

Now that we have seen how much prime source energy goes into global electricity production, we will look at other end of the electrical grid to see how it is consumed by the major sectors. As was shown in the electricity flow chart, Figure 5-2, of the 189 QBTU used for global electricity generation, only 63 QBTU (33 percent) actually reaches end-use consumers.

Figure 5-5 gives the predicted growth of consumed electricity out to the year 2030.The figure shows the demand for electricity to be increasing by more than two percent yearly. In comparison, the world's human population is growing at about one percent yearly. It has been estimated that there are nearly two billion people in the world who still do not have access to electricity. They want it. Can you blame them? Also the demand for more electricity from those who already have it (us, for example) only compounds the problem. Faced with this situation, it is difficult to see how any headway will be made on reducing global electricity consumption anytime in the near future.

## Section 5: Energy on the Home Front

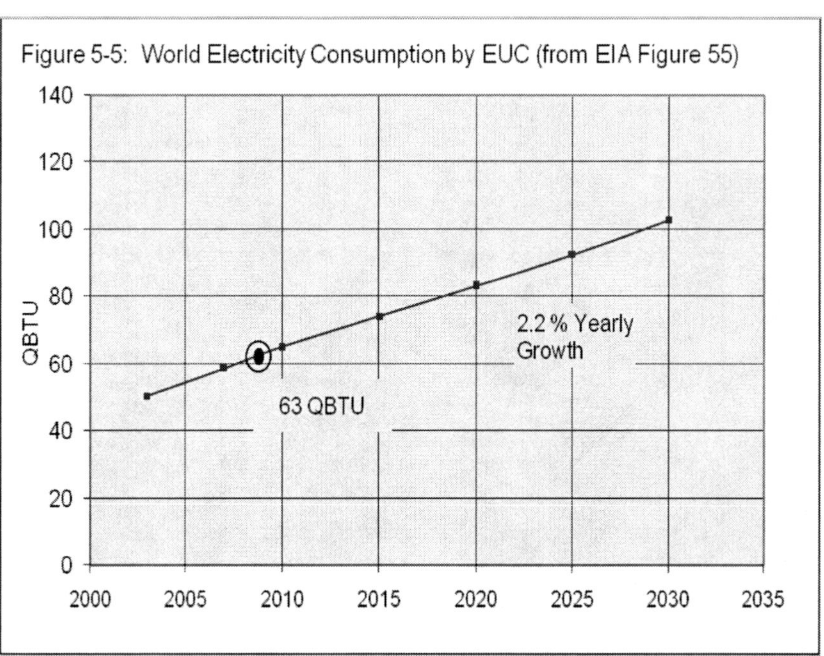


### Electricity Consumption by the Sectors

Figure 5-6 shows where electricity ends up in terms of the four major consumption sectors. The electricity distribution among the sectors is: industry 46%, residential 29%, commercial 22%, and transportation 3%. As you may recall from Section 3, the breakdown for total end-use energy consumption within the major sectors, Figure 3-4, was somewhat different: industrial 50%, transportation 27%, residential 15% and commercial 8%.

This shift in energy distribution between the sectors reflects the very small amount of electricity, less than 3 percent, used by the transportation industry. It is noted that while electricity consumption for the industrial, residential and commercial sectors is predicted to grow at a steady upward pace, electricity for transportation is predicted to remain essentially flat through 2030. This is discouraging news for both the electric vehicle and the rapid transit industries.

## Section 5: Energy on the Home Front

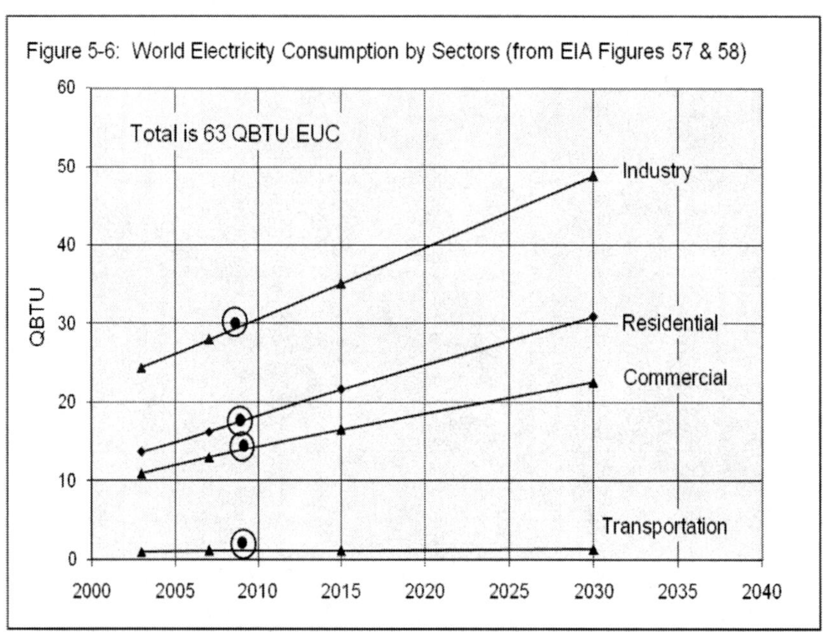


> Electricity use will to continue grow in the industrial, residential and commercial sectors; in the transportation sector, however, electricity use will remain flat.

*A Useful Table*

Here is a table you might find useful when dealing with electrical power and energy units. When reading about electricity, any one or more of these units may be encountered (often in the same article), so be prepared.

| Table 5-2: Electricity Units and Abbreviations | | | | |
|---|---|---|---|---|
| Unit | Abbr. | Unit | Abbr. | Factor |
| watts | W | watt-hour | Wh | 1 |
| kilowatts | kW | kilowatt-hour | kWh | 1E+03 |
| megawatts | MW | megawatt-hour | MWh | 1E+06 |
| gigawatts | GW | gigawatt-hour | GWh | 1E+09 |
| terawatts | TW | terawatt-hour | TWh | 1E+12 |

## Section 5: Energy on the Home Front

### What Does All This Really Cost?

Let us assume that the electrical energy consumption for a typical household is about 10,000 kilowatt hours. At a cost of $0.15 per kilowatt hour (typical residential rate including taxes etc.) this works out to be about $1,500 per year. If you live in a mega-size home and an area of extreme summer weather conditions, your yearly bill will be much higher than this typical value. But if you are a typical user, this amounts to about $4 on a daily basis, less than the cost of a fast-food burger with fries (or a gallon of gas). Such a deal.

*Wait, It Gets Better*

But wait, that $4 is based on your total electric bill. Don't forget that your bill includes the operations and maintenance (O&M) costs of your utilities company plus various local taxes in addition to the actual cost of the electricity. Think of all the costs required to keep your electricity flowing: linemen, heavy equipment, meter readers, finance and management personnel and so on.

We can make a rough estimate of electrical utilities O&M costs by considering our phone or high-speed Internet bills, which are at least $30 to $50 a month for basic service. These are primarily O&M costs since there is no actual "product" being delivered. We can assume then that about $40 per month (which is probably low) goes into your electricity bill for O&M and tax purposes.

This reduces the cost of the actual electricity to about $1000 per year, which works out to an unbelievable $2.80 per day! Think of it, $2.80 a day for the electricity required to run everything listed in Table 5-1. No, you won't believe it – you will still complain about "skyrocketing" electricity costs. This seems to be part of our gene structure.

Next we will look at electricity costs adjusted for the Consumer Price Index (inflation) over the past several decades to see how much rates have really increased.

## Section 5: Energy on the Home Front

*Residential Electricity Historic Rates*

Figure 5-7 shows how typical residential electricity rates have fared relative to the general inflation rate. You won't believe this either – everyone knows electricity rates have been skyrocketing. On the contrary, electricity rates have closely followed the general inflation rate. To be fair, there are many varied and complex energy-rate formulas for big-time industrial and commercial users which are not considered here and which may not necessarily follow this trend.

There are many versions of this chart available on the Internet. They are all somewhat different in specific values, but they all tell the same story. You can make up your own chart, as I did, if you are a non-believer. First, look up the historical electricity rates (dollars per kilowatt-hour) for the region of the country where you live. Then use one of the many Consumer Price Index sites found on the Web to adjust then-year-dollar electricity rates to current-year dollars. Your actual values may be a little different than those given here, but the shape of the curves should be very similar. My chart says electricity rates, when adjusted for inflation, have actually been on a decline over the past twenty-five years or so. So much for skyrocketing electricity rates.

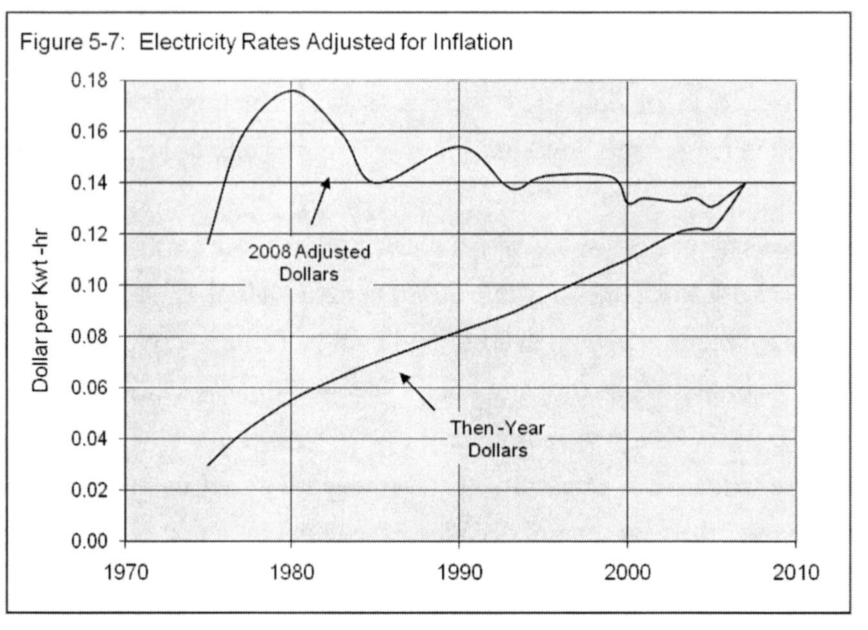

Figure 5-7: Electricity Rates Adjusted for Inflation

## Section 5: Energy on the Home Front

*Natural Gas Historic Rates*

The story for natural gas rates is a little different than for electricity. Figure 5-8 shows a big jump in rates from the mid-seventies to the mid-eighties. After 1985 natural gas rates leveled off until about 2000 when they took a pronounced jump. From the mid 1980s to 2000, when the rates were falling relative to inflation, I do not remember seeing then any newspaper articles proclaiming that the cost of natural gas was "plummeting". The point is we have been on a cheap ride with natural gas for the last twenty years or more, and maybe the ride is over. There was a time, you may remember, when excess natural gas was considered a nuisance and safety hazard in oil fields and was routinely burned off. No one knows how many QBTU of world energy were lost because of this practice.

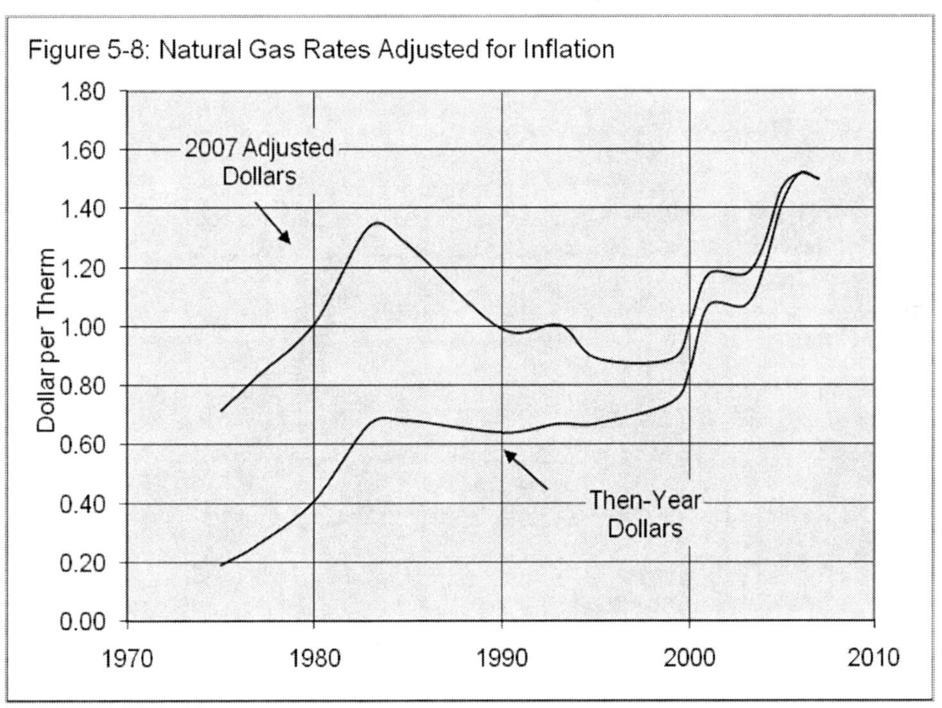

# Section 5: Energy on the Home Front

## Residential Energy Consumption by Item

Figure 5-9 shows typical household energy consumption as the percentage of world EUC. The figure includes all forms of household energy; electricity, natural gas, fuel oil, coal, propane, butane and biofuels. Since there is no exact definition of a "typical" global household, the figure represents a composite based mostly on U.S. residential consumption, but with some world residential data mixed in. Overall the figure probably best represents U.S. and Western European household energy consumption.

Household heating (aka space heating) is the major player in residential energy consumption – approximately 55 percent of this sector. This, in turn, is equivalent to 8 percent of global end-use consumption (EUC). Appliances plus other household electrical devices account for about 2.4 percent of EUC. Water heating consumes about 2 percent. Air conditioners, clothes dryers and lighting come in each at one percent or less. These are broken out separately to highlight their contribution to global energy consumption; they often are pointed out as examples of where homeowners could save *significantly* if they were more energy conscious (i.e., less wasteful).

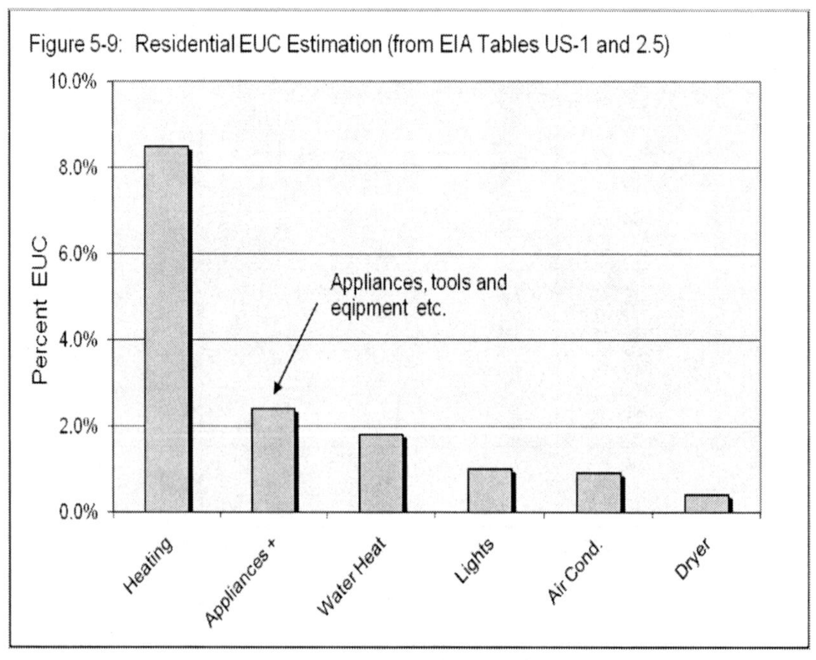

Figure 5-9: Residential EUC Estimation (from EIA Tables US-1 and 2.5)

## Section 5: Energy on the Home Front

*Average U.S. Household Direct Energy Use*

In Figure 5-10 the left bar shows how much energy the typical U.S. family uses energy directly in terms of their heating, electricity and gasoline bills.

The yearly household energy for electricity was estimated on 10,000 kilowatt-hours EUC, which would be the equivalent of approximately 30,000 kilowatt-hours PSE. The energy for household heating was based on 800 therms per year of natural gas.

The energy for gasoline was based on18,000 miles per year between two vehicles getting 20 miles per gallon each. Together the utilities and gasoline add up to 295 MBTU yearly of direct energy for the typical U.S. family.

Next we will look at the "real" or total energy consumed by the typical U.S. family. When paying your utility bill or buying gasoline, you are well aware of the cost of energy. All of the other products and services that we use, however, have large built-in energy costs. Normally we don't think much about this energy – someone else is using it, not us.

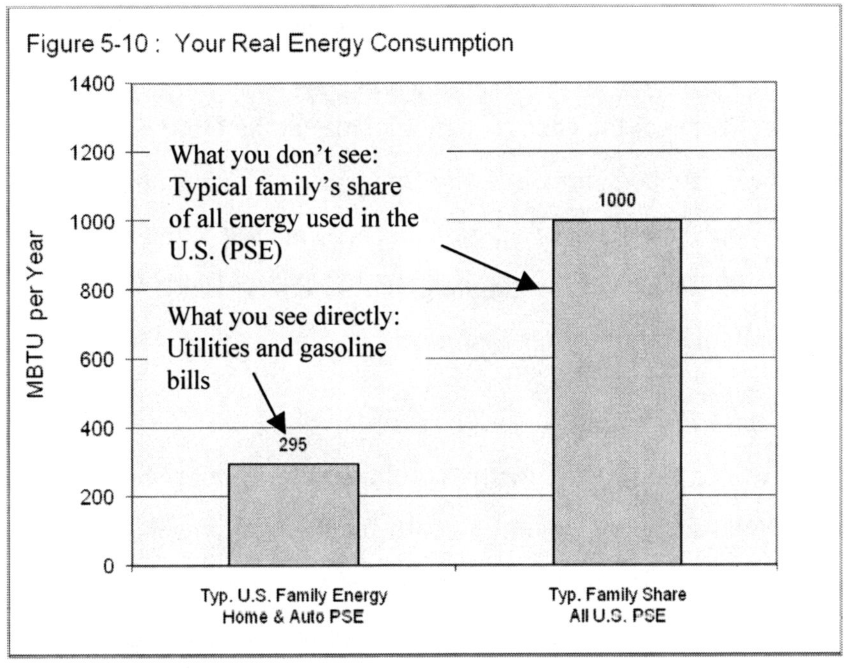

## Section 5: Energy on the Home Front

*Average U.S. Household "Real" Energy Use*

The total or "real" energy consumed by a typical U.S. family is shown in Figure 5-10 (right bar). This real energy was estimated by dividing total U.S. prime source energy (about 110 QBTU) by the number of households in the U.S. (about 110 million). The people at the U.S. census bureau may take exception to my definition of the "typical family" and to my estimation techniques, but the message is clear – the typical U.S. family uses about three times more energy overall than it uses for utilities and gasoline alone. This is the energy associated with consumer purchases as well as the energy shared indirectly with the other EUC sectors. I know what you are thinking: "We don't use all of that other energy; someone else is using it – the big corporations, industry, and the government." Yes, they are using it, but they are using it on your behalf as a consumer of goods and services.

**Household Energy Use – The Global Picture**

We have seen how the U.S. family spends energy directly for home and automobile use. Now let's look at how this compares with the energy spent by households at the global level. Here a "top down" approach will be used. In Figure 5-11(next page) the energy spent globally at the family level (household and automobile) has been broken away from the other consumption sectors. It is seen that about 31 percent of all global energy is used at the family level. This agrees with the previous estimate given for the typical U.S. family which was done using the "bottoms up" approach.

*The Residential Flow Chain*

In Figure 5-12 the flow of energy is followed from initial production down to the residential sector and ending with the energy for household lighting. The figure shows why switching to fluorescent bulbs, something we should all do, is not as effective for reducing fossil-fuel consumption as we have been led to believe – saving only 1.4% of all PSE even if everyone in the world switched.

## Section 5: Energy on the Home Front

Figure 5-11: Energy Use at the Global Level

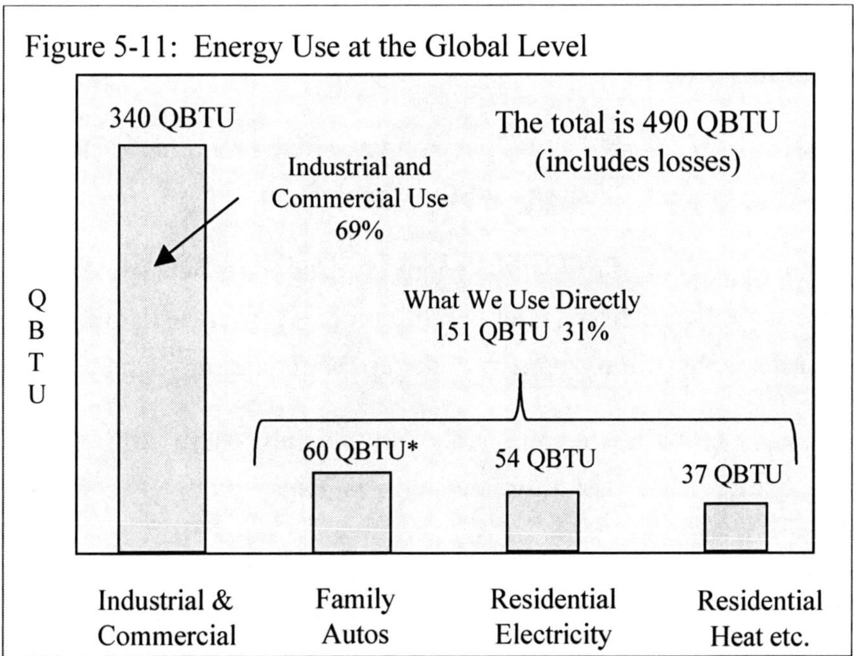


* Estimated at 80% of all gasoline consumption

Figure 5-12: The Residential Energy-Flow Chain

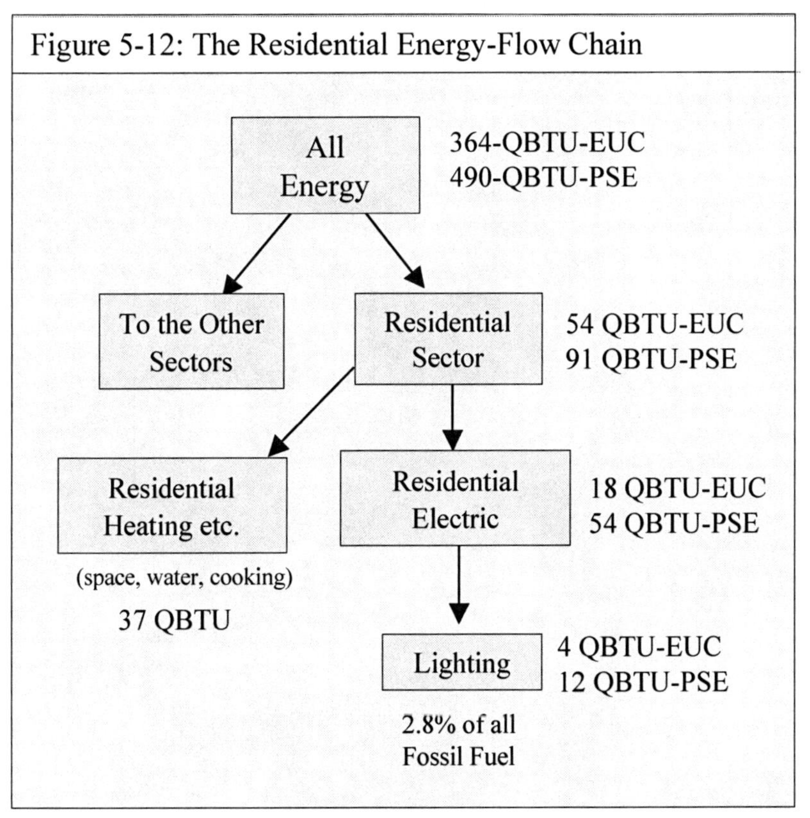

# Section 5: Energy on the Home Front

## Summary of Section 5

The electrical energy that we use at home (EUC) requires at least three times that amount of energy (PSE) to generate it.

About 55 percent of residential energy is used for room (space) heating. Appliances, water heaters, air conditioning, electronics, lighting and other items that are plugged-in make up the remaining 45 percent.

Although we all like to complain about the high cost of utilities, when adjusted for inflation, utility rates have remained relatively constant over the past 25 years.

Energy consumption is embedded in everything we do and buy. The average U.S. family uses about three times more energy than it pays for directly through utilities and gasoline bills.

# Section 6

# Energy and Your Automobile

Passenger vehicles consume 12 percent of all global energy (60 out of 490 QBTU), 32 percent of all oil (60 out of 185 QBTU) and 14 percent of all fossil-fuel energy (60 out of 429 QBTU). As such, the automobile receives a high degree of attention from the media and the environmental community as a major energy-user and contributor to greenhouse emissions. These attention-grabbing issues will not be addressed here. Instead we will look at more mundane things like miles per BTU and miles per dollar for various vehicle and fuel types. We will look at the cost of gasoline over the past decades relative to the general inflation rate. Also we will estimate the amount of energy required to run your car for a year as compared to feeding your body for a year. Finally, energy efficiency comparisons between gasoline, hybrid, electric, hydrogen and biofuel vehicles will be made.

## Vehicle Fuel Efficiency – Miles per BTU

The subject of miles per BTU (fuel-efficiency ratio) for the various motor vehicle types can be very controversial. Everyone wants their favorite form of automotive transportation to be seen in the best light. The vehicle parameters assumed for the comparisons made in this section – vehicle size, weight, range, fuel type – can often be more important than the actual vehicle type itself. For example a smaller, efficient gasoline vehicle can have a more favorable efficiency ratio than a larger-sized hybrid vehicle.

But even more important is the consideration of EUC versus PSE energy. If nothing else, this section will have you asking whether they are talking about *end-use energy* or *prime source energy* when encountering claims about automobile efficiency.

## Section 6: Energy and Your Automobile

*Biofuel Vehicles*

As you know there is the ongoing debate over how much energy it takes to produce a gallon of ethanol. Ethanol production factors include the energy for farm equipment, fertilizers and the water resources to grow the corn (or other biosource material). Then there is the energy required to actually produce and deliver the ethanol to the consumer. In Section 8 (Alternative Energy Sources) these issues for ethanol will be looked at in further detail. Here we will only cover the efficiency of ethanol-powered vehicles as compared to the other vehicle types.

*Electric Vehicles*

For electric vehicles there is the issue of how much prime energy is required to generate the electricity that ultimately reaches the vehicle's battery through the home (or other) outlet. Electric vehicles themselves are very efficient users of energy, but as we learned in Section 3, for every BTU that comes out of your home electrical outlet about three times that amount of energy was required to produce it. This would be an important factor if there were to be a large-scale global switch to electric vehicles. Initially, coal and natural gas would provide the first-line source of additional energy needed for all of these new electric vehicles. At least this would be the case until a significant number of solar, wind and other alternative energy sources came on line to help carry the load.

*Fuel-Cell Vehicles*

Like electric vehicles, fuel-cell vehicles claim the promise of reduced dependence on imported oil along with reduced greenhouse emissions. This is true only when hydrogen end-use energy is being considered. The important factor here is the energy required to transform prime source energy (coal or natural gas) into the hydrogen used by the fuel-cell vehicle. In each step of the hydrogen-making process (as you will see), energy losses take their toll, thereby reducing the overall PSE efficiency of the vehicle.

## Section 6: Energy and Your Automobile

### Calculating Vehicle Efficiency – Two Approaches

The PSE approach for assessing auto efficiency is referred to as the "cradle to grave" or "well-to-wheel" approach. This approach accounts for the energy to produce the fuel as well as for the energy used by the vehicle itself.

You may encounter auto efficiencies based on the EUC approach where terms like "tank-to-wheel" or "battery-to-wheel" are used. More often than not, it may not be stated which approach is being used. The tank-to-wheel approach is useful for making direct comparisons between particular vehicle types, but does not address the real issue: how much prime source energy is effectively lost before reaching the vehicle's tank or battery?

The PSE approach addresses this issue. It begins at the oil well, the coal mine, the natural gas field and at the nuclear or hydroelectric plant.

The parameters used to make the vehicle-efficiency comparisons are summarized in Tables 6-1, 6-2a/b and 6-3. For the gasoline vehicles, average and high-mileage versions were considered. It is noted that an apples-to-apples comparison was attempted. The vehicle parameters were selected to represent, as best as possible, typical mid-size vehicles rather than smaller, but obviously more energy-efficient vehicles that you often hear about.

*Miles per Million BTU – By EUC*

When EUC is used as the basis for vehicle efficiency, the lithium-ion electric vehicle heads the list as shown in Figure 6-1. Although this provides good PR for electric vehicles, this approach doesn't give the whole energy story for the reasons indicated above. Next on the EUC efficiency list is the hydrogen fuel-cell vehicle. Both of these vehicles attest to the efficiency of electricity for doing useful work – the fuel cell being essentially a battery that powers the electric motor of the hydrogen-fuel vehicle.

## Section 6: Energy and Your Automobile

Note that the remaining vehicles on the EUC efficiency chart all have internal combustion engines using either gasoline or ethanol. Missing from the chart for now is the hybrid plug-in vehicle that you may have been reading about. Until reliable data on energy consumption under both "gasoline power" and "electric power" becomes available, their EUC and PSE efficiencies will remain in question.

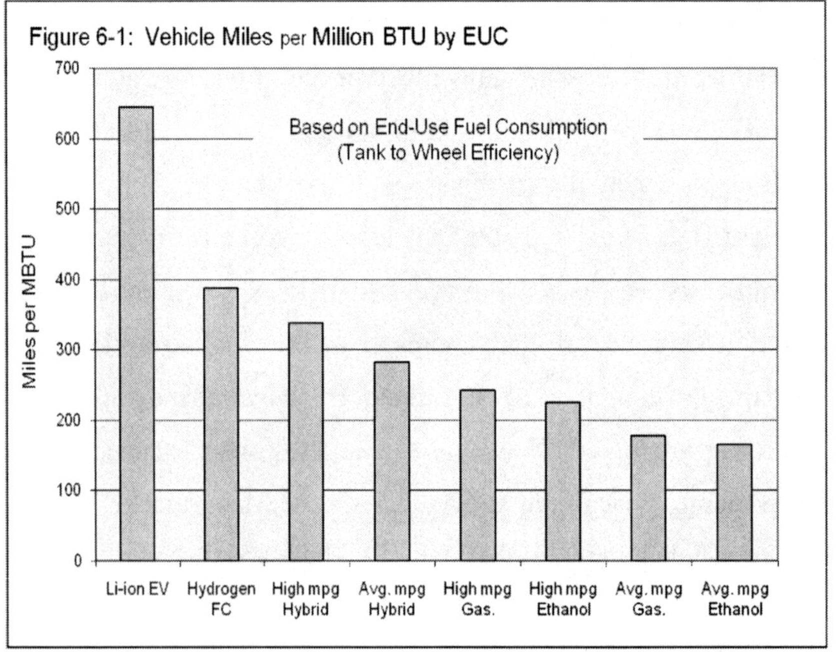


*Miles per Million BTU – By PSE*

Figure 6-2 (next page) shows the PSE (well to wheel) efficiencies for the same set of vehicles. It is immediately noted that these efficiency rankings are quite different from the EUC rankings given in Figure 6-1. Ethanol vehicles now head the efficiency list when PSE is used as the basis for comparison (more to be said about this later).This figure also shows that PSE efficiencies can vary as much as a factor of two between the vehicle types (150 to over 300 MBTU per mile).

## Section 6: Energy and Your Automobile

Hydrogen-fuel vehicles, because of the many energy-losing steps inherent in hydrogen production, now fall into the mid-range of PSE efficiency.

Electric vehicles (lithium-ion battery) fall close to gasoline vehicles in PSE efficiency. This is probably contrary to everything you have been hearing about electric vehicles. Much of what you have heard is based on EUC efficiency or the assumption that the electricity for EVs will come from solar and wind energy sources. For these comparisons it is assumed the electricity will come from the electrical grid using existing conventional fuel sources.

The effective energy losses that occur when going from prime energy (PSE) to the fuel tank or battery (EUC) for each of the vehicles will be shown in Figure 6-4 in a series of step-by-step diagrams.

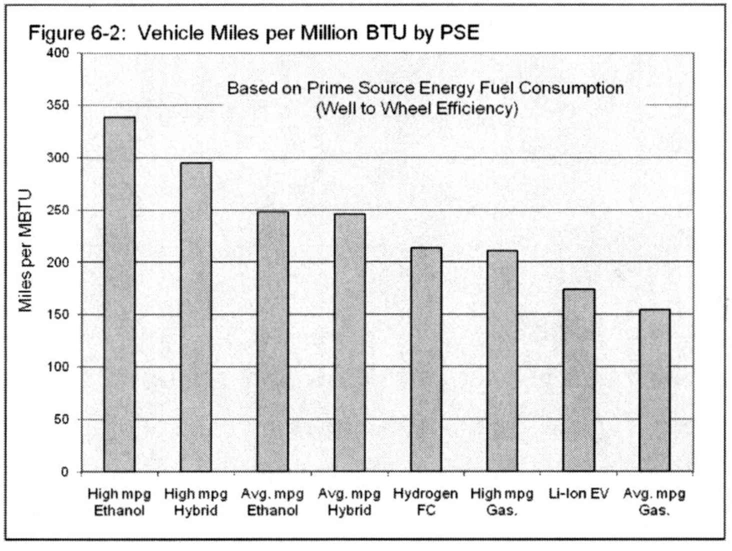


> Ethanol and gasoline-hybrid vehicles are the most energy-efficient from a prime source energy (PSE) standpoint.

Next we will take a closer look at the parameters used for the vehicle efficiency rankings and also at the assumptions made for the production of ethanol, electricity and hydrogen.

## Section 6: Energy and Your Automobile

*Fuel Vehicle Parameters*

Table 6-1 lists typical parameters for gasoline and ethanol internal combustion engine vehicles. The ethanol vehicle mileage is based on ethanol fuel having about two-thirds the energy equivalent of a gallon of gasoline. For each vehicle type, both average and high-mileage versions were considered. All versions were assumed to be driven 14,000 miles per year. The dollars per gallon reflect national average costs as of early 2008. By the time you read this, fuel costs may be considerably higher, but the relative costs per mile for the vehicle types should remain nearly constant.

Table 6-1: Fuel Vehicle Parameters "Tank to Wheel" EUC

| Vehicle Type > | Avg. mpg Gasoline | High mpg Gasoline | Avg. mpg Hybrid | High mpg Hybrid | Avg. mpg Ethanol | High mpg Ethanol |
|---|---|---|---|---|---|---|
| Miles per gallon | 22 | 30 | 35 | 42 | 12.5 | 17.1 |
| BTU per gallon | 1.24E+05 | 1.24E+05 | 1.24E+05 | 1.24E+05 | 7.60E+04 | 7.60E+04 |
| Miles per BTU | 1.77E-04 | 2.42E-04 | 2.82E-04 | 3.39E-04 | 1.65E-04 | 2.25E-04 |
| BTU per mile | 5636 | 4133 | 3543 | 2952 | 6061 | 4444 |
| **Miles per MBTU*** | **177** | **242** | **282** | **339** | **165** | **225** |
| Dollars per gallon | 4.00 | 4.00 | 4.00 | 4.00 | 3.75 | 3.75 |
| Cents per mile | 18.2 | 13.3 | 11.4 | 9.5 | 29.9 | 21.9 |

* "Tank to Wheel " EUC Efficiency

*Electric Vehicle Parameters*

Tables 6-2a and 6-2b give the parameters used for making the electric vehicle (EV) comparisons. The tables are presented in both English and metric units separately to clear up the confusion that often occurs when electric vehicle data is given in mixed units. The specific energy values (BTU/lb or kWh/kg) for the batteries are typical for modern EVs, although you may hear about advanced batteries that promise much higher storage capacity. The estimated driving ranges are for typical driving conditions, not for steady-state highway driving, which is always unrealistically high. The electricity rate of 0.15 dollars per kilowatt-hour is typical for Northern California as of early 2008. The parameters represent, as best possible, mid-sized EV's comparable to those listed in Table 6-1 for the gasoline and ethanol-fueled vehicles.

## Section 6: Energy and Your Automobile

| Table 6-2: Electric Vehicle Parameters "Battery to Wheel" EUC | | | |
|---|---|---|---|
| English Units | Lithium-ion | Ni-Metal Hydride | Super Lithium-ion |
| Battery BTU per lb | 233 | 124 | 310 |
| Battery weight, lbs | 800 | 800 | 800 |
| Battery charge, BTU | 1.86E+05 | 9.91E+04 | 2.48E+05 |
| Vehicle range, miles | 120 | 60 | 200 |
| BTU per mile | 1.55E+03 | 1.65E+03 | 1.24E+03 |
| **EUC Eff. *, miles per MBTU** | 645 | 605 | 807 |
| Electric rate, $ per kWh | 0.15 | 0.15 | 0.15 |
| Electic rate, $ per MBTU | 44 | 44 | 44 |
| Cents per mile | 6.8 | 7.3 | 5.4 |
| Metric Units | | | |
| Battery kWh per kg | 0.15 | 0.08 | 0.2 |
| Battery weight, kg | 363 | 363 | 363 |
| Battery charge, kWh | 54 | 29 | 73 |
| Vehicle range, miles | 120 | 60 | 200 |
| kWh per mile | 0.454 | 0.484 | 0.363 |
| **EUC Eff.*, mile per kWh** | 2.2 | 2.1 | 2.8 |
| Electric rate, $ per kWh | 0.15 | 0.15 | 0.15 |
| Cents per mile | 6.8 | 7.3 | 5.4 |
| * "Battery to Wheel" EUC Efficiency | | | |

*The Super EV*

Very high efficiencies for electric vehicles are often found in the literature. These, while being numerically correct, can be very misleading. This is how it's done. First select a small, very lightweight EV, the kind you see in the newspaper under the heading "Electric Vehicle Gets 300 Miles per Charge." Usually they look like a car a cartoon character should be driving. Then use the steady-state value for driving range instead of the more realistic everyday driving range. Finally, use end-use energy consumption (not prime source energy) to compute the vehicle's energy efficiency. By using this approach efficiencies of over 1,000 miles per MBTU can be calculated. It could also be claimed that this vehicle was pollution-free and did not create greenhouse emissions. This would not be correct since the energy required to generate the electricity for the vehicle (mostly from fossil fuels) has not been taken into account.

## Section 6: Energy and Your Automobile

*Electric Vehicle Batteries*

Batteries have been steadily improving in energy storage capacity over the last several decades. Battery storage capacity can be given as watt-hours per kilogram (energy) or watts per kilogram (power). In Figure 6-3a the storage capacity of four EV battery types is shown on a BTU-per-pound basis.

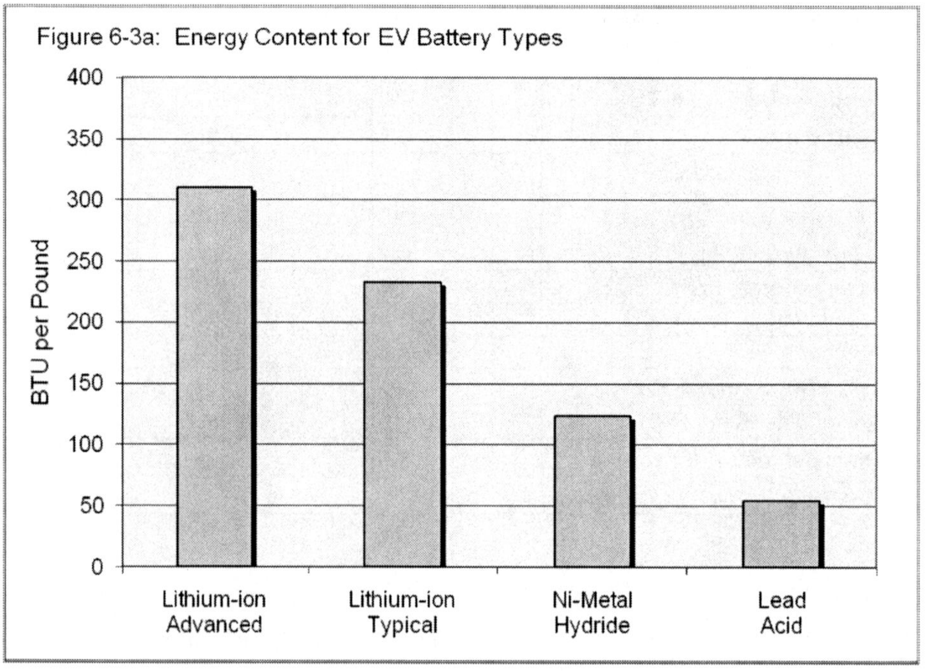


This figure shows the progression of battery storage capacity from the lead acid battery, around for decades, to the latest lithium-ion battery that is used to power the newest electric vehicles as well as computers and cell phones. As great as these improvements have been, battery storage capacity per unit weight still falls far short from that of gasoline and the other petroleum-derived fuels as will be seen in the next figure.

## Section 6: Energy and Your Automobile

*Battery to Gasoline Comparison*

In Figure 6-3b the storage capacity (BTU per pound) for the batteries shown in the previous figure are compared to gasoline. Even with the great strides being made in battery technology, the storage capacity of even the most advanced lithium-ion batteries still remains some seventy to a hundred times less than that of the equivalent weight of gasoline.

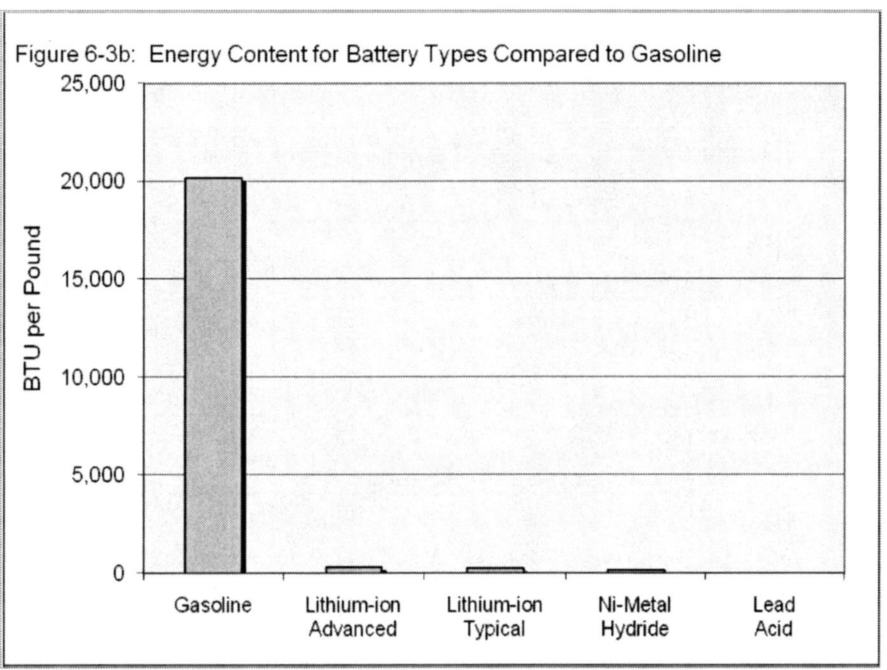


*Hydrogen Fuel-Cell Parameters*

Table 6-3 gives the parameters that were used for the hydrogen fuel-cell (HFC) vehicle-efficiency calculations. Since these vehicles are not yet a mass-produced consumer item, the parameters are tentative, to say the least. They were compiled from the limited number of technical, operational and promotional descriptions for hydrogen fuel-cell vehicles that are available on the Internet.

## Section 6: Energy and Your Automobile

The calculation of fuel-efficiency for a hydrogen vehicle is a simple matter. The only information needed is the storage capacity of the hydrogen tank (in pounds or kilograms) and knowledge of how far the vehicle can travel on that tank of hydrogen. The real issue is how much PSE energy is required to produce the hydrogen fuel. This is more problematical than estimating the prime source energy needed for an electric vehicle. At each step in the production of hydrogen, there are losses that are dependent on the technologies being used. The steps in the generation of hydrogen for a fuel-cell vehicle are shown diagrammatically in Figure 6-4.

| **Table 6-3: $H_2$ Vehicle "Tank to Wheel" EUC** | |
|---|---|
| Tank capacity, lbs $H_2$ | 10 |
| $H_2$ BTU per lb | 5.67E+04 |
| Tank capacity, BTU | 5.67E+05 |
| Vehicle range, miles | 220 |
| Miles per lb $H_2$ | 22.0 |
| Miles per BTU | 3.9E-04 |
| **Miles per MBTU*** | **388** |
| * "Tank to Wheel" EUC Efficiency | |

*Other Vehicle Efficiency Studies*

There are many vehicle-efficiency comparisons to be found on the Internet and in the literature. Many are technically complex, using things like gearbox efficiency, aerodynamic drag and tire-to-road friction as parameters. Some writers attempt to consider both EUC and PSE in their efficiency comparisons; however definitions of PSE can become a little misleading. In one study on electric vehicles, the calculated efficiency appeared to be inordinately high. The reason for this was that the author used wind power as the prime energy source instead of the existing electrical grid as should have been used. Others have used solar as their prime energy source. These are examples of how the "time tunnel" can be used to make a favorite vehicle appear more energy-efficient than it really is.

## Section 6: Energy and Your Automobile

One of the most useful vehicle-efficiency studies is from Kreith and West (Ref. 14). Their definition of vehicle types is somewhat different, but overall their results match up fairly well with those given here.

### Auto Vehicle Efficiency Chains

Figure 6-4 gives my version of the *Auto Vehicle Efficiency Chains* for the several vehicle types under consideration. The starting point of the chain is one million BTU of prime source energy.

*How the Vehicle Efficiency Chains Work*

The efficiency chains show the steps involved in getting prime source energy to the vehicle's tank or battery. At each step along the way there are inevitable energy losses. Upon reaching the fuel tank (or battery), the vehicle's EUC efficiency factor is used to calculate the miles that can be driven on the remaining amount of effective PSE energy.

The energy losses for each step in the chain are typical and very approximate. Critics will always be able to point out an advanced or future technology which could reduce losses in any particular step. The point of the table is to give you a sense for the number of steps and resultant energy losses that occur before the prime energy source reaches the vehicle.

*From Well to Tank to Wheel*

The production and distribution of gasoline is a relatively efficient process with estimates falling in the 87 percent range. Both conventional gasoline and hybrid vehicles benefit from this in terms of their PSE efficiencies.

## Section 6: Energy and Your Automobile

*From Power Plant to Battery to Wheel*

The efficiency of the electrical grid to generate and deliver electricity to end-use consumers is about 33 percent as we have seen from the energy-flow chain (Figure 3-7). Currently (no pun intended) the electrical grid is the only viable way to provide energy for electric vehicles at the global level. Yes, you could recharge your EV from solar cells on your roof [(1)], but for now this a personal solution, not a near-term global solution.

The electric vehicle's "battery-to-wheel" EUC efficiency of 645 miles per MBTU is the highest of any vehicle in the efficiency chain. However when combined with the 33 percent electrical grid efficiency, the overall PSE efficiency (power plant to battery to wheel) drops to 174 miles per MBTU.

*From Field to Tank to Wheel*

PSE efficiency for the ethanol vehicle remains highly controversial in that it depends on the energy balance assumed for the growing and production of ethanol. In this work the USDA ratio of 1 to 1.67 (input to output energy) is used in the ethanol vehicle efficiency calculation. From the diagram it is seen that one MBTU of input energy results in 1.67 MBTU available for vehicle use. This boost in energy from the sun gives the ethanol vehicle its relatively high PSE efficiency.

*From Natural Gas to Tank to Wheel*

The production of hydrogen for fuel-cell vehicles involves more steps than any other in the chain diagrams – from natural gas (or coal), to hydrogen gas, to liquefied hydrogen, to the vehicle's tank with storage losses, to the fuel cell for electricity generation, to the electric motor and finally to the wheels of the vehicle. Despite these energy losses, hydrogen fuel-cell vehicles still show a reasonable PSE efficiency – this due to their relatively high EUC efficiency of 388 miles per MBTU.

---

(1) Assuming you are home during the day and it is not cloudy or overcast.

## Section 6: Energy and Your Automobile

**Figure 6-4: Auto Vehicle PSE Efficiency Chains**

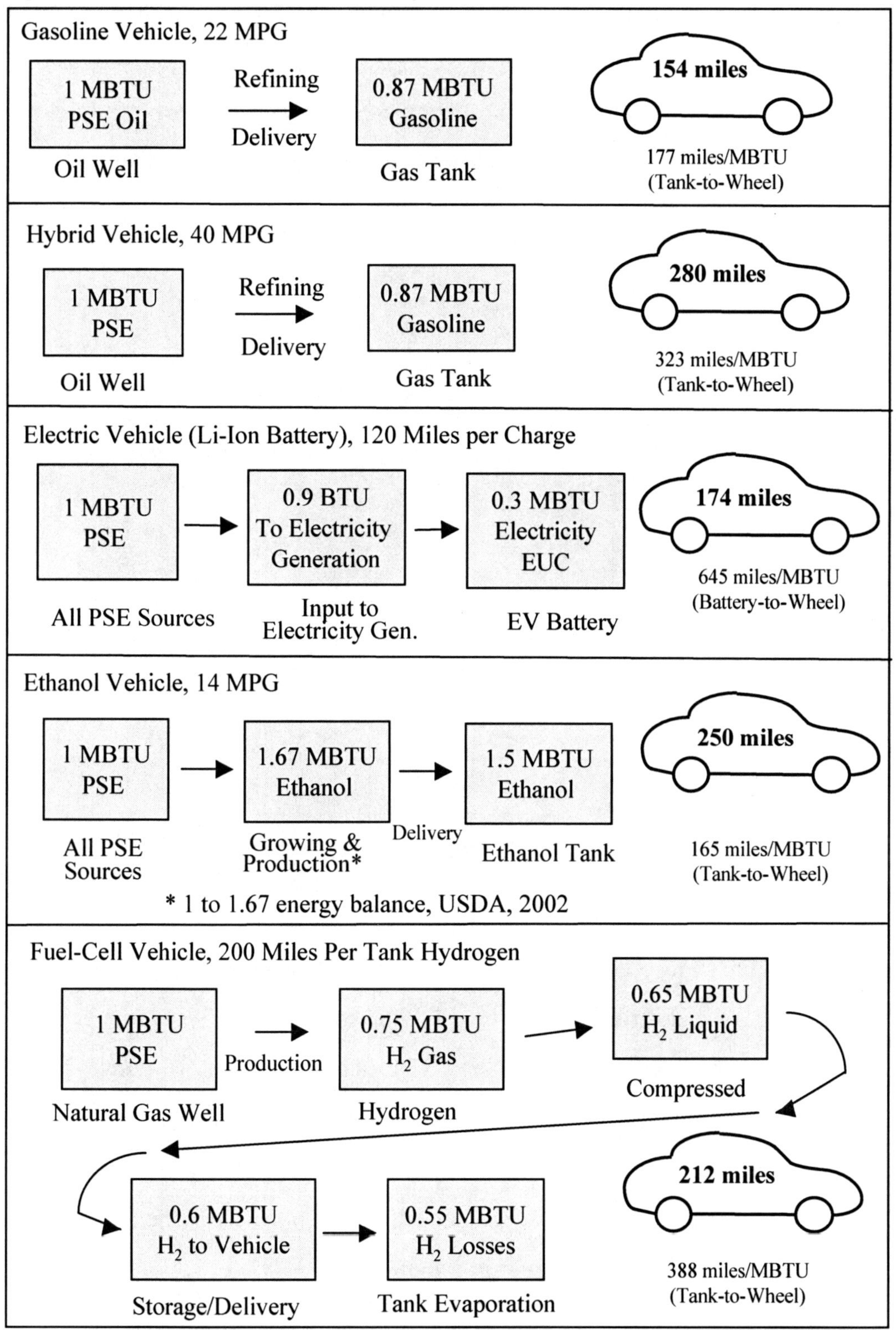

## Section 6: Energy and Your Automobile

### Energy Cost by the Mile

Normally when we think of driving costs, we think of the cost of gasoline and the mileage we get from our car. Figure 6-5 compares the cost of driving in terms of cents per mile for each of the vehicle types we have been considering.

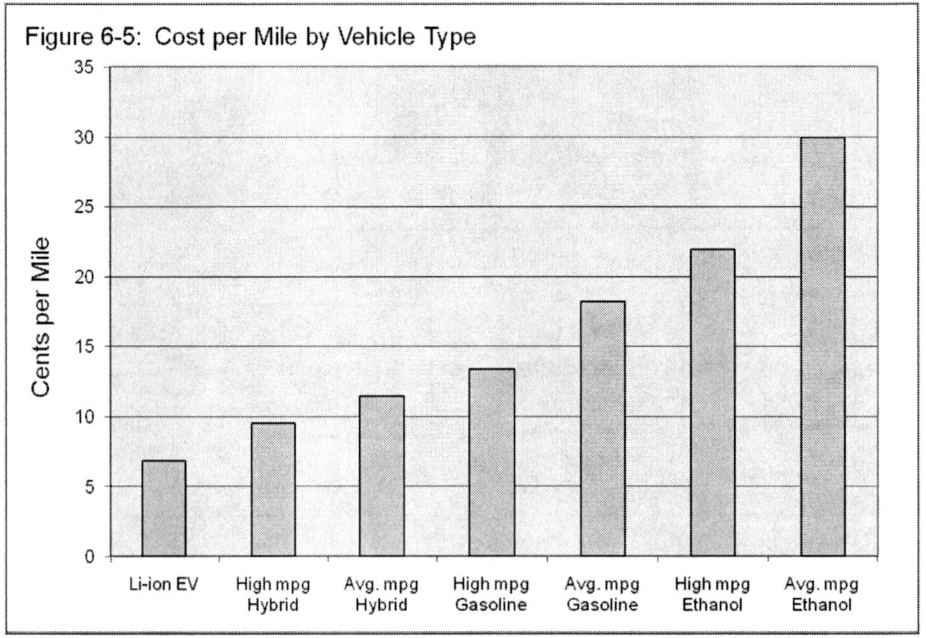


> On an energy-cost-per mile basis, an electric vehicle with a lithium-ion battery is the least expensive to drive.

This figure shows that the least expensive vehicle to drive, based on fuel cost alone, is the electric vehicle with a lithium-ion battery. This is true even though this vehicle scores lower in the PSE efficiency category (Figure 6-2). This points out the need to question all claims on vehicle-efficiency. Is it the cost-per-mile efficiency or is it the energy-per-mile efficiency that is being addressed? A vehicle with a low cost per mile is not necessarily the best vehicle from the prime source energy standpoint.

## Section 6: Energy and Your Automobile

*Your Car vs. Your Body*

Figure 6-6 compares the energy used by your car with the energy used by your body (your food). These estimates, although very approximate, serve to make the point how much energy is devoted to our automobiles.

The energy to run your car is based on a 22-mpg vehicle being driven for 12,000 miles yearly. For your body, 2,000 large Calories per day were assumed. Estimates for the energy to produce the food you eat can vary by factors of three or four depending on how far up the food chain you go. The food chain begins with farm equipment, fertilizers, insecticides, irrigation and initial processing and then goes on to commercial processing, distribution and retailing. See Section 3, Agriculture Sub-sector, for the assumptions made here.

The energy required to build a car is difficult to pin down. Eight percent of the manufacturing and resources sub-sectors were assumed here. This would be considered low by many since it is not a "cradle to grave" estimate; this would include maintenance and other operating costs as well as disposal costs.

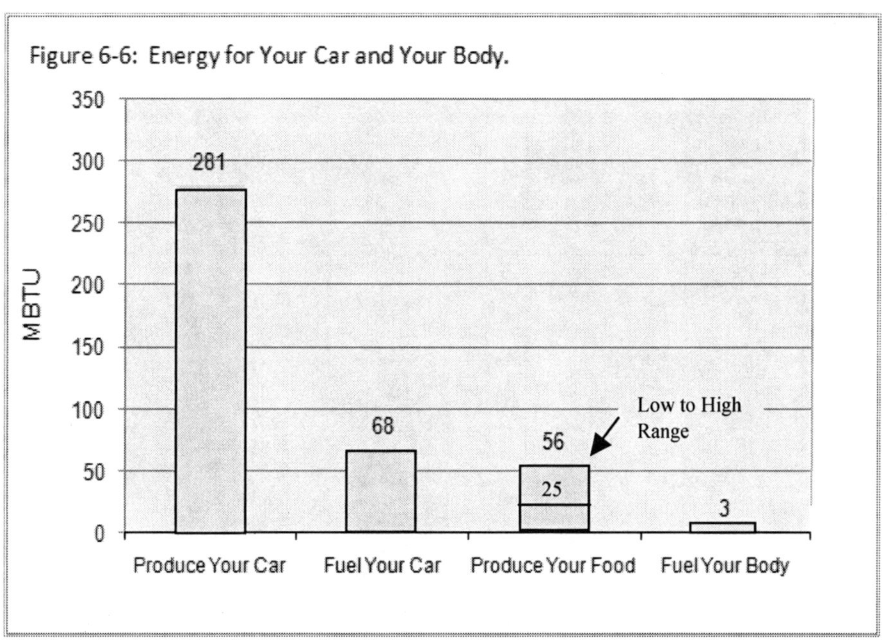

Figure 6-6: Energy for Your Car and Your Body.

The energy to run your car is more than the energy to grow your food for a year. Globally speaking, a car is like another mouth to feed.

## Section 6: Energy and Your Automobile

*Cost of Gasoline Adjusted for Inflation*

Petroleum and gasoline are experiencing their greatest cost increases since the energy crises of the late 1970s. I know this won't make you feel better or help your pocketbook, but let's see how gasoline prices look when adjusted for inflation over the past fifty years. As was the case for the inflation-adjusted electricity and natural gas rates given in Section 5, if you don't believe these results you can make your own. The historic cost of gasoline can be found on numerous Websites. Consumer inflation-rate factors are also easily found. Combine these and you will get a figure that looks pretty much like Figure 6-7. The figure points out two important things. First, gasoline prices are now rising at an unprecedented rate ("skyrocketing," as the news media proclaims). Second, gasoline prices have now exceeded the all-time peak of the 1980s "Energy Crisis" years. We have had an easy ride cost-wise since 1980 considering the increasing global demand, greenhouse emissions and the other detrimental environmental effects associated with gasoline. It appears the ride has ended.

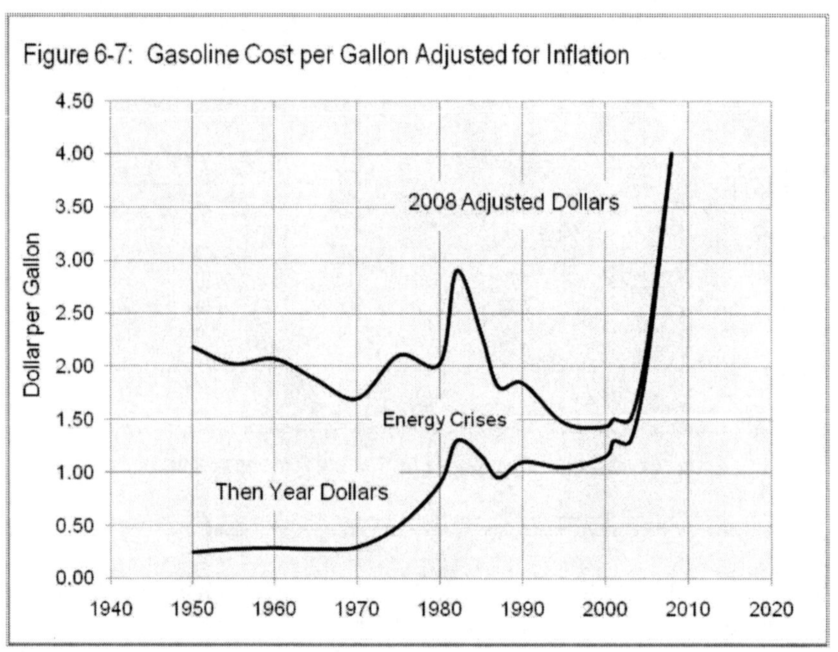

Figure 6-7: Gasoline Cost per Gallon Adjusted for Inflation

## Section 6: Energy and Your Automobile

*Cost of Driving Adjusted for Inflation and Increased Mileage*

You have probably seen versions of Figure 6-7 in newspaper or magazine articles. It is usually given as an incidental footnote to the real issue of "skyrocketing gasoline prices."

There is another factor to the price of gasoline that is also rarely mentioned. This is the factor of improved mileage of the average automobile. Since the mid-1960s average mileage has risen from about 12 mpg to 22 mpg (see Figure 6-9). When the cost-savings from improved mileage are combined with the inflation-adjusted cost of gasoline, the result is Figure 6-8. What does the figure say? It says that on a cost-mileage adjusted basis, the cost to drive a mile today is about the same as it was way back in 1960, even with gasoline over $4.00 per gallon. I know what you're thinking, "Who cares what gasoline costs back then, I only know it costs me big bucks to fill my gas tank today." Well, what more can I say?

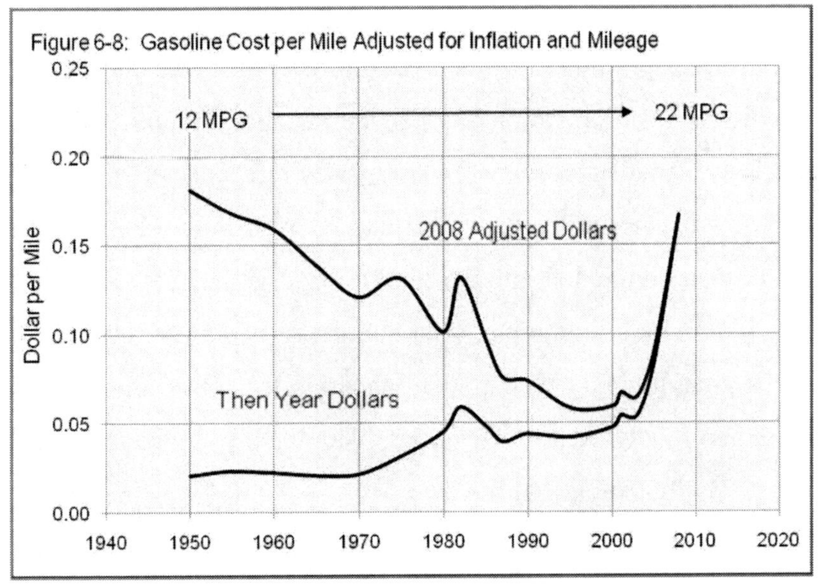


> When adjusted for inflation and improved gas mileage, the cost to drive per mile has changed little since the mid-1960s.

## Section 6: Energy and Your Automobile

### Thoughts on Mileage

Mileage is a conditional property of any motor vehicle, not a completely designed-in property. The mileage estimates used in this section are based on average or typical driving conditions and thus they are neither "highway" nor exactly "city" mileage. What is typical? Well, it's not the steady-state, constant-speed mileage you would get on an Interstate on a Sunday morning. This would be closer to "highway" mileage. It's not the mileage you would get on this same highway during the speed-up then slow-down commute period. It's not the mileage you would get trying to drive through Manhattan on a bad traffic day; this is would be "sub-city" mileage. Typical mileage, as used here, is probably most closely represented by the "city" value given on the window sticker of new automobiles.

*Real-Life Everyday Mileage*

We all like to talk about our great gas mileage. We usually give anecdotal or personal-best mileage figures when talking to friends – like our last vacation trip down the Interstate to visit relatives when we got 32 mpg. This probably represents only a small percentage of our yearly miles driven, but it sounds good. To know the real mileage of your car, keep track of all gasoline used for a year or so, then divide that into the total miles driven. This would cover everything – your daily commute, your countless trips around town doing whatever you had to do, as well as your vacation trips. You may not like the answer you get. It is likely to be even less than the city-mileage estimate that came on the window-sticker of the car.

# Section 6: Energy and Your Automobile

## Thoughts on Improved Mileage

Auto mileage, at least for the gasoline engine, has probably reached the point of diminishing returns with the advent of the hybrid vehicle. Further increases in gas mileage will occur, but they will be marginal. Unless a completely new type of gasoline engine is developed, the best we will ever see will probably be in the range of 40 to 50 mpg (average, not highway). You will often hear of an automobile that gets fantastic mileage, like 70 or 80 mpg, but these are always very lightweight and small vehicles that most of us would not want to drive (but may have to someday). Figure 6-9 shows the history of improved mileage for auto vehicles since 1965. Note that the average vehicle mileage nearly doubled from 1965 to 1980, but has remained essentially flat since then. The mileage data spread shown in the figure has to do with the definition of passenger cars and light trucks and "on the road vehicles" versus "newly sold vehicles" – none of this all that important for our purpose here.

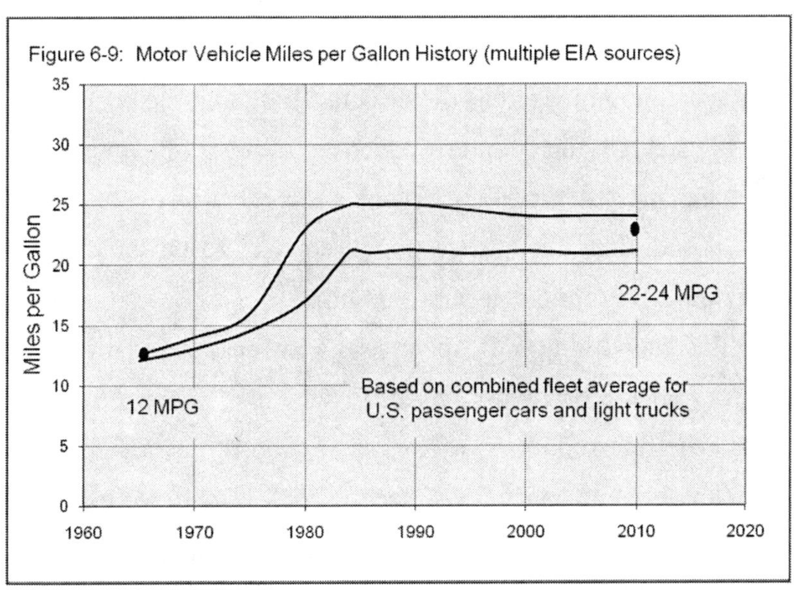


> Gasoline mileage has doubled since 1965, but the number of vehicles in the world has quadrupled.

## Section 6: Energy and Your Automobile

*The Hybrid Plug-In*

The hybrid plug-in automobile runs on a combination of a gasoline and electricity. It is somewhat like a hybrid vehicle, but the battery is larger and it can be recharged from the electrical grid. For short-range trips around town, the car would run mostly on electricity. For longer trips or when the battery was low, it would run on gasoline. You may have heard that hybrid plug-ins might achieve over 100 miles per gallon. This an unfair claim because it ignores the electrical energy used to attain the miles driven. If the "under gasoline power only" mileage were given, it would probably be less than for an equivalent-size regular gasoline vehicle due to the additional battery weight the car must carry around. For honest comparisons, hybrid plug-in vehicles will need to have their mileage expressed in the appropriate energy terms which account for both the electrical and gasoline energy used. The hybrid plug-in does not exist as a consumer vehicle at the time of this writing. We will just have to wait and see how energy-efficient these vehicles actually turn out to be.

*Heresy, Heresy*

Increasing gasoline mileage has the effect of reducing the cost per mile driven. This may actually be counterproductive from a global energy-savings perspective. What, are you crazy? Everyone knows that this is one of the most widely touted proposals for saving gasoline, reducing fossil-fuel consumption and for saving the environment at the same time.

Think of it this way, higher mileage means a lower cost per mile. This will be an incentive for many to drive more miles – "why not, I have a fuel-efficient car." This has been referred to as the *rebound effect*. It may also be an incentive for a family with only one car to buy a second car – "the driving cost will be low, we can afford it." It may also be a factor for many to take a job with a longer commute. Is this just idle speculation? Let's look at what has been happening with the world's auto population, the average number of miles driven per year and the consumption of gasoline over the last thirty years.

## Section 6: Energy and Your Automobile

We have seen (from Figure 6-9) that average auto mileage has risen from about 12 mpg in 1970 to 22 mpg in 2007. We also know that the average miles driven per year has increased from about 10,000 in 1980 to over 13,000 today as shown in Figure 6-10. Do people have more places to go and things to do than they did in 1980 or is this the *rebound effect* of improved fuel economy?

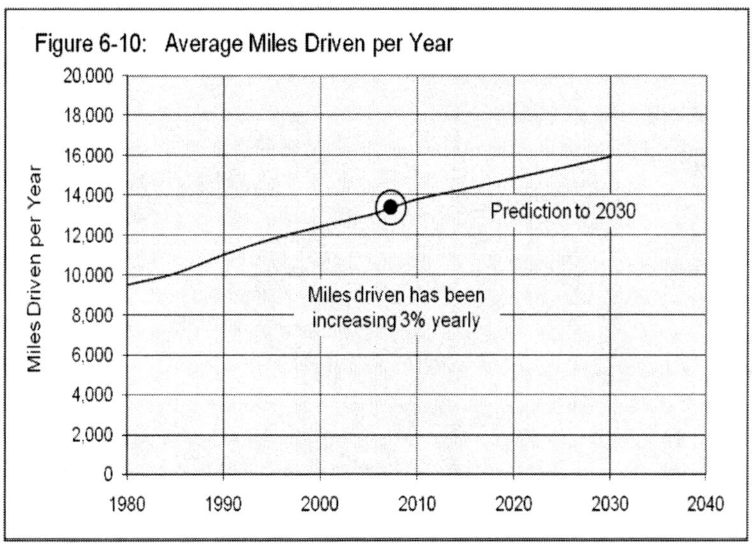


We know that yearly gasoline consumption has quadrupled from 150 billion gallons in 1960 to over 600 billion gallons yearly in 2007 (Figure 6-11).

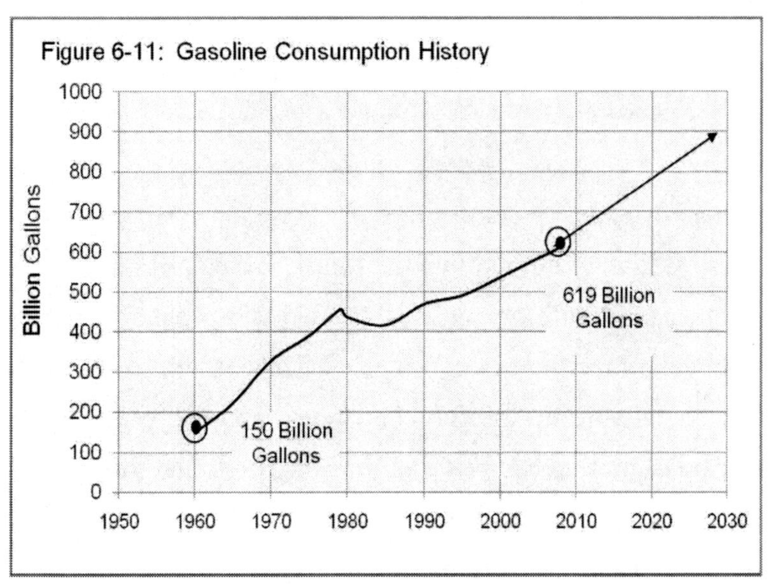

## Section 6: Energy and Your Automobile

And finally we know from Figure 6-12 that the world's motor vehicle population is increasing at the rate of between 2 to 3 percent yearly. The real counterproductive effects of improved fuel efficiency will be from the developing economies. Millions, who in the past couldn't afford to buy one, will soon be able to own a motor vehicle. India's new Tata vehicle, the Nano, will be very fuel-efficient, and at about $2,500 each you can be sure that millions will be sold worldwide in the upcoming decades.

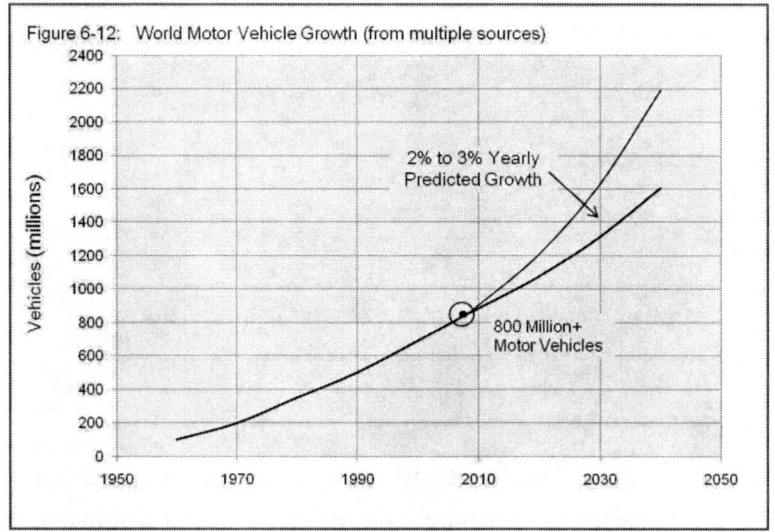


Figure 6-12: World Motor Vehicle Growth (from multiple sources)

The world's motor vehicle population is increasing faster than the world's human population

The net effect of the factors just mentioned appears to have washed away the gasoline savings from past increases in fuel efficiency. There is little reason to think that these factors will not dilute the future fuel savings hoped for from hybrids and other high-mileage vehicles. Someone is probably doing research on this now to determine the extent of these effects on future gasoline consumption. A study of increased fuel economy and the resultant effect on U.S. gasoline consumption can be found in Ref. 16. A first-cut estimate of gasoline and fossil-fuel savings from a global switch-over to hybrid vehicles is given in Section 9 (How to Use This Handbook).

## Section 6: Energy and Your Automobile

*The Petroleum Energy Flow Chain*

Figure 6-13 tracks the flow of fossil-fuel and petroleum consumption down to the motor-vehicle level. Motor vehicles use approximately 80 percent of all gasoline, 32 percent of all petroleum, and 14 percent of all fossil fuel. In Section 9 this diagram will be used to show how this flow of fuel consumption would be affected by future high-efficiency gasoline vehicles and also by future electric vehicles.

Figure 6-13: The Petroleum Energy Flow Chain

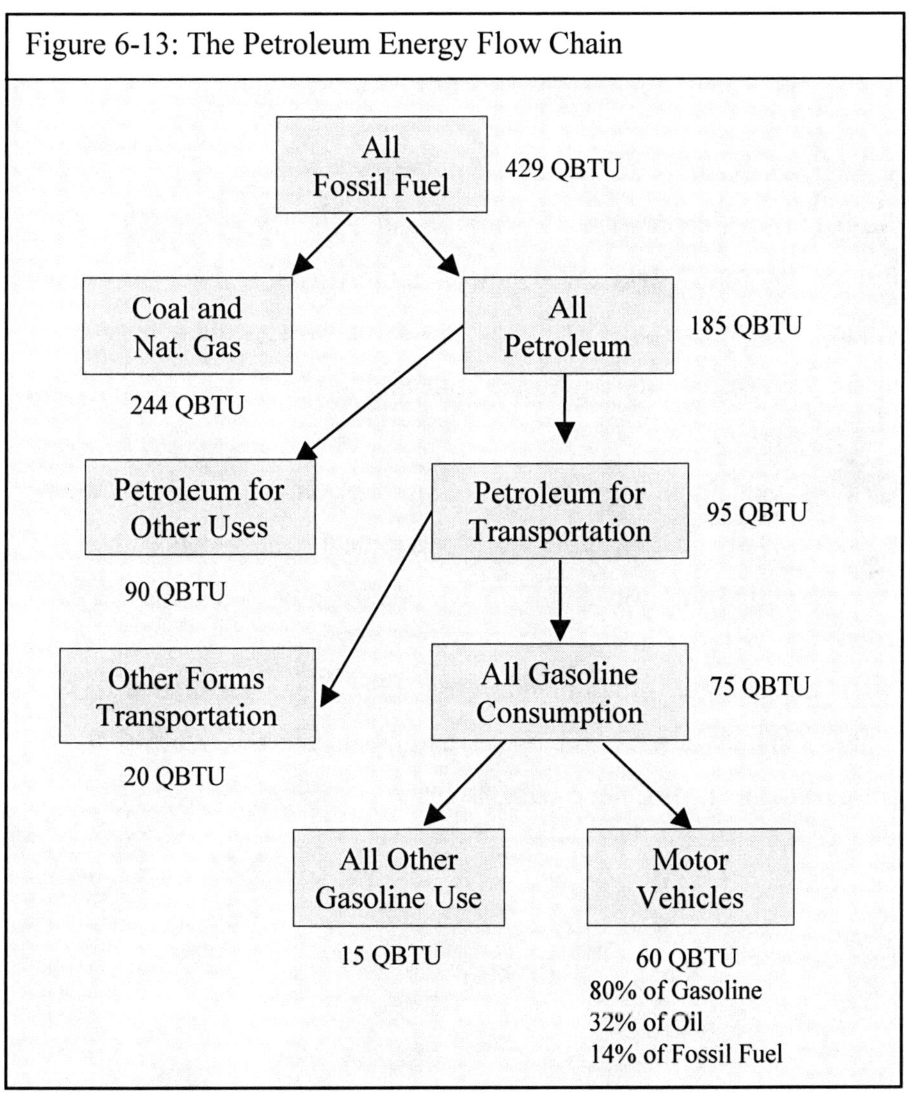

## Section 6: Energy and Your Automobile

### Summary of Section 6

Electric and hydrogen fuel-cell vehicles appear to be the most energy efficient when looked at from the EUC viewpoint. However, when looked using the PSE approach, these vehicles show energy efficiencies similar to the higher mileage gasoline vehicles.

Ethanol vehicles are the most energy efficient from the PSE standpoint, this based on the USDA energy production ratio for ethanol of 1 to 1.67 (many researchers take exception to this).

The claims made by advocates of any energy-saving vehicle, electric or otherwise, should always be questioned. Are they using EUC or PSE energy to calculate efficiency? Are they using realistic everyday mileage? Are they considering the inevitable worldwide increase of motor vehicles on future gasoline consumption?

Gasoline vehicle mileage has almost doubled since the early 1970s, but the number of vehicles in the world has almost quadrupled. This has resulted in gasoline consumption now being almost twice that of 1970.

When adjusted for inflation and improved mileage, gasoline costs per mile are about the same as they were in the mid-1960s. At over $4.00 per gallon this is difficult to accept, but that's the way it is.

# Section 7
# Looking Back at Energy Savings

In this section we will look at the many energy-savings initiatives introduced over the last three decades to see their effect on reducing global energy consumption. Figure 7-1 shows the growth in global energy use since 1970. At the continued rate of two percent yearly, global energy consumption is predicted to reach 700 QBTU annually by the year 2030.

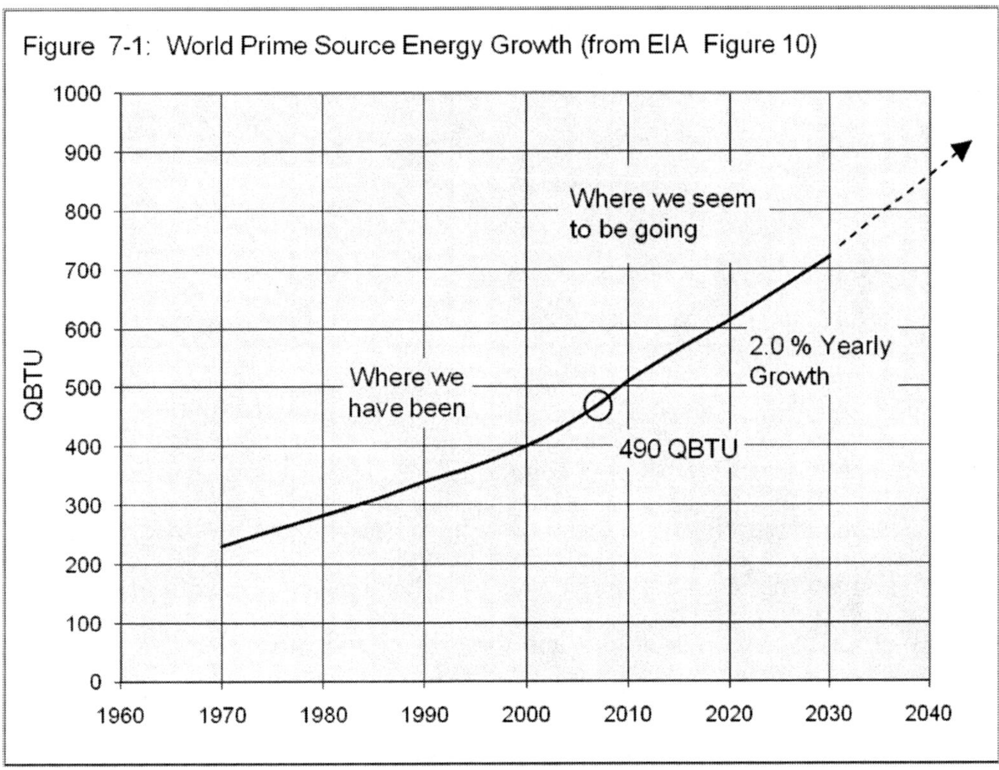

Figure 7-1: World Prime Source Energy Growth (from EIA Figure 10)

## Energy Wake-Up Calls

The 1973 oil embargo and the Iran-Iraq War were energy wake-up calls for the entire world. Since then numerous energy conservation and efficiency initiatives have been proposed. In Table 7-1 summarizes the more well known of these. Thirty-five years later these same items still routinely appear in the media as ways to save energy and the environment.

## Section 7: Looking Back at Energy Savings

| Table 7-1: Energy-Saving Initiatives Since 1970 |
|---|
| Since the Energy Crunch of the 1970s: |
| Auto mileage has increased significantly. |
| Energy-efficient appliances have been introduced. |
| Building codes revised for improved energy efficiency. |
| Public awareness increased on energy issues. |
| Energy-Saving Ideas: |
| Drive less, car pool. |
| Lower thermostat settings. |
| Use fluorescent bulbs. |
| Install double-pane windows. |
| Buy a fuel-efficient car. |
| Use clothesline instead of dryer. |
| Recycle, recycle, recycle. |

Automobile average mileage has almost doubled since 1970. Energy-efficient appliances have become commonplace. The conversion efficiencies for producing electricity from fossil fuels have been improved. Local, state, federal and international building codes have been put in place requiring more energy-efficient homes, buildings and factories. Public awareness on energy issues has been greatly increased.

We have been urged to drive less, to car pool and to buy fuel-efficient cars. We have even been asked to keep our auto tires properly inflated. We have been told to reset our thermostats, replace incandescent bulbs with fluorescent, install double-pane windows and use a clothesline instead of our clothes dryers. Recycling has become part of our daily lives. According to some reports, we are getting two to three times more useful work out of the energy we now consume (the data on this varies greatly). Despite all of this, global energy consumption continues on its relentless climb.

# Section 7: Looking Back at Energy Savings

## How Much Has All of This Helped?

Looking at Figure 7-1 again, it appears that all of these energy-saving initiatives have not had much effect on reducing or even slowing down global energy consumption. OK, without them current energy consumption would certainly be even higher than it is now. In any case, thinking that the continued application of these conservation measures will somehow result in significant declines in future energy consumption doesn't seem likely.

Why haven't these initiatives added up to more significant global energy savings? They have not because:

### *The Global-Savings Shrink Factor*

Energy savings at the global level will always be a *percentage-times-percentage-times-percentage* thing. This means energy savings get very small very quickly when looked at from the global level. Numerical examples of the shrink factor in action will be given in Section 9.

Here is a notional example of how of the *Global Shrink Factor* works. Home lighting is about 10 percent or more of residential electricity depending on household usage. This would appear to be an area of great potential savings, but residential electricity is only about 28% of all electricity consumption (18 out of 63 QBTU). Electricity, in turn, is only about 17 percent of all end-use energy (63 out of 364 QBTU). String these factors together and you arrive at the realization that home lighting makes up only about one percent of all global energy consumption (EUC). So even if everyone in the world cut their home lighting by one-half by going to fluorescent bulbs, the result would be a small reduction of global energy consumption.

### *Global Energy Demand Never Sleeps*

Energy savings are up against a heavyweight contender – *Global Energy Demand.* To appreciate global energy demand, you must again refer to Figure 7-1.

## Section 7: Looking Back at Energy Savings

The demand for global energy appears to be insatiable. It is fueled by six billion-plus humans relentlessly chewing away at the world's energy resources. Global energy consumption makes the world go around. It is the foundation of the global food chain as well as the basis of the entire global economy – at least in the way it operates today. Energy demand, consumer spending, the gross world product and greenhouse emissions, as you probably suspect, are closely intertwined. In Section 10 (The Energy Triad) you will see simplified diagrams and graphs showing how the elements of the Triad are numerically related. You don't have to be a Nobel-prize winner to get the picture.

*While You Save, the World Keeps Turning*

True, you use a lot of energy when driving your car, but think of all the energy it took to process the raw materials, to build the hundreds of parts and sub-assemblies, to assemble it, to market it, to ship it and to retail it at your local car dealer. Don't forget the bank that gave you the loan and the insurance company that insures it. They all use energy on your behalf. While you are trying to conserve energy, these mega-energy consumers keep grinding away, day and night.

*It's More Than You and Your Car*

Savings initiatives for most of us fall in the residential and automobile sub-sectors of energy usage (see Table 3-1). These sectors account for about a third of the world's energy consumption. The world's remaining energy is consumed by the industrial, construction, business, commercial and agricultural sectors as was seen in Figure 3-9. Although we may think of the industrial and the other big energy-consuming sectors as beyond us or as having nothing with us, they in fact are us. Every purchase we make or every service provided to us is based on the energy expended by these big-time energy users. Why would they spend all of that energy if not ultimately for us, the consumer?

## Section 7: Looking Back at Energy Savings

*The Let's-All-Work-Together Approach*

Finally, most energy-saving proposals, as just pointed out, fall into the residential and personal auto categories. These savings proposals are largely based on voluntary conservation. This comes under the category of *It just isn't going to happen.* The "Hey gang, if we all work together we can do this" approach works only in old Hollywood movies. Look at Table 7-1 again to remind yourself of the many innovative and seemingly sure-thing energy-saving initiatives that have been instituted since the early 1970s. It will be very difficult to get people to save even more energy. Why? Because most people don't think they are wasting energy, they are just doing what they have to do to get by in the world they live in.

### What, Me Waste Energy?

Everyone, including you, believes that they are not wasting energy. Talk to Mr. or Mrs. Anyone about energy savings, and you will get a story somewhat similar to this:

*"We use our cars for getting back and forth to work, for personal business and occasionally for vacations. With gasoline prices so high, why would we do unnecessary driving? We set our thermostat to $68^0$ F in the winter and to $80^0$ F in the summer. We had to buy new appliances a few years back and they were all energy-efficient. We save up our laundry so that we only do full loads in the washer and dryer. We have already installed many fluorescent bulbs in our house. With utility rates increasing so fast, why would we waste energy? Get real."*

And so on and so on. I know what you are thinking; you could find all sorts of places where the Mr. and Mrs. could save more energy. But it doesn't matter what you or I think, the point is that the Mr. or Mrs. think that they are not wasting energy.

## Section 7: Looking Back at Energy Savings

### What Is Energy Waste?

Everyone has a different view on what energy waste is (sounds just like government spending, doesn't it). Here is an exaggerated example: you see a soft-drink truck making deliveries. There must be hundreds of thousands or even millions of these things around the world burning up gasoline while delivering soda water. Is this energy waste or not? If you are a drinker of the product, the truck driver or a soft-drink executive, the answer is no way, this is not energy waste.

On the other hand if you are a nutritionist/environmentalist type, you see waste; nothing but flavored sugar water, something no one really needs, something that in the long run is bad for your health, not to mention your overloaded bathroom scale.

You have probably heard this a thousand times: "We are all energy addicts. We must stop wasting so much energy if we are going to solve this problem." If only things worked this way. If you encounter someone who actually thinks he or she is wasting energy, please let me know.

### Summary of Section 7

The many energy-saving initiatives introduced since the 1970s have done little to reduce global energy consumption or even slow down the growth rate.

Energy savings are difficult to achieve because most energy consumption is considered useful and necessary – at least by the users.

If there is one big myth about energy use, it is that large amounts are being wasted and "if only we weren't so wasteful," we could solve this problem.

Saving energy at home and in our automobiles, while a noble effort is outmatched by the prodigious and ever-growing amounts of energy expended in the industrial, manufacturing, agricultural and commercial sectors.

# Section 8: The Alternative Energy Sources
# (Old Fossils Need Not Apply)

As we have seen earlier, the alternative energy sources aren't doing very well against the old fossils. Do you remember why? The fossil fuels have a strong starting lineup which makes them very difficult to compete against:

***High energy-concentration – Greatest bang for the buck.***
***Transportability – Can be delivered to virtually anywhere in the world.***
***Storage capability – Can be stored anywhere in the world, indefinitely.***
***Full-time worker – No down-time, always ready to perform.***
***Readily Available – Even with the threat of dwindling supplies.***

In this section we will look at the limiting factors behind solar, wind and biofuel energy sources. You will see that these limiting factors are not technological in nature. There already exist an endless number of polemic studies to be found on the Internet describing why this or that alternative energy source is not "technically or economically" feasible. It is assumed here that the issues of efficiency, cost, reliability, safety and producibility can and will eventually be solved. We will look here at the more fundamental limitations.

## Near-Term vs. Long-Term Energy Solutions

Before discussing the inherent limitations of the alternative energy sources, a reminder on near-term versus long-term solutions is appropriate. In the context of this work, near-term solutions are those which are feasible within the next twenty-five to thirty years when the double threats of reduced oil supplies and greenhouse emissions reach the point of no return.

In discussions with proponents of the various alternative energy sources, any limitations mentioned will immediately be countered with "time tunnel escape hatch" solutions. These solutions, though plausible for the future, are just not going to happen in the near term.

## Section 8: The Alternative Energy Sources

As an example, you are in a discussion about electric vehicles and their potential wide-scale use. You point out that the energy for all of these electric cars will have to come, at least in the near term, from the increased use of fossil fuel for electricity generation. The electric-vehicle proponent will say that we can get all of the electricity we need by simply building a vast network of solar (or wind or nuclear or geothermal) energy facilities. OK, but near term? You get the idea; watch out for the "time tunnel escape hatch" when future energy solutions are being discussed.

### Solar Energy Generation

What are the limiting factors for solar energy generation? Is it the efficiency for converting the sun's energy to useable energy? After fifty years of steady improvements, solar cell efficiencies are still only in the 12 to 15 percent range (laboratory panels have reached 50 percent efficiency). On a yearly average, even at this low efficiency, today's solar systems can still generate significant amounts of energy; enough to pay your yearly electricity bill in many cases.

Is it the cost per kilowatt-hour? For now yes, but you can find numerous studies on the Internet which show future costs per kilowatt-hour favorably approaching the cost of fossil fuel generated electricity (as the cost of these increases).

Are solar systems too expensive to install? Maybe for a lot of people, but the costs are coming down and in most cases are subsidized by federal and state agencies. On a daily basis you can read about large corporations that have gone solar. So if efficiency and cost are not the limiting factors, what are they?

*Solar Energy Limiting Factors*

The limiting factors for solar energy generation are:

1. Solar systems produce only electricity and electricity is less than 40 percent of global PSE (189 out of 490 QBTU).

## Section 8: The Alternative Energy Sources

2. Solar systems are part-time workers. For about 75 percent of the time (based on 365 days/24 hours), solar systems do not produce effective energy: night, dawn, late afternoon, cloudy and overcast days etc. When not generating electricity, all solar systems must be backed up with conventional sources of electricity from your utilities company.

3. There is no viable way to store the excess electrical energy generated when solar systems are producing at peak levels.

What do these inherent limitations mean in terms of solar energy becoming a major source of global energy production? What is the most that solar energy can do for us?

*From the energy-flow chain, about 133 QBTU of fossil fuel is used for global electricity generation. Solar systems could replace maybe 30 percent of this or about 40 QBTU. The fossil-fuel savings would be about 10 percent (40 out of 429 QBTU). This amount of savings could not be exceeded no matter how efficient solar systems became or how many facilities were built.*

*The More the Better?*

That doesn't make sense. Everyone knows that the more solar facilities we build, the more energy we will have. No, it's not a matter of how much electrical energy we can generate; it's a matter of how much electricity we can use on a real-time basis. Once the maximum electrical grid demand is met, extra solar capability will only serve as backup or idle capability.

Let's say you lived in a remote, rugged area. You need a gasoline powered generator to supply electricity for your cabin. You determine that a two-kilowatt generator will give you all the electricity you need even on the coldest and darkest winter nights. You buy one. Would you then buy another generator?

## Section 8: The Alternative Energy Sources

Maybe, but only as a backup in case the first one went out. But then would you buy another, then another and yet another after that? Of course not because you would realize that the additional capability would be of no value, you already have all of the electricity you can use from the first generator.

*Residential Solar Systems*

If you install an appropriate-size solar system and you live in a relatively sunny region, the system can pay for your entire annual electric bill. This is because the excess energy generated during bright sunny days runs your electric meter in reverse allowing you in effect to "sell" energy back to your utilities company.

*A typical home solar system might be characterized as follows:*

*4 kilowatt system at a cost of $25,000 to $30,000 after rebates;*

*9,000 kilowatt-hours of generated electricity yearly;*

*$1,350 saved on your yearly electricity bill (at $0.15 per kWh)*

This system will pay back the installation cost in about 20 years or even less if electricity rates continue to increase. *If you take out a home equity loan or make a withdrawal from your retirement account to pay for the system be sure to subtract this from your estimated annual electricity savings*. You may find the loan payment or the lost earnings exceed the cost of the electricity you think you are saving.

Also remember that solar systems are feasible today only because they are backed up by your local utilities company using energy mostly from, guess what, coal or natural gas. It would be pretty cold and dark around your place on winter nights if you had to depend completely on your solar system. If only you could store up all that excess energy produced during those sunny summer days.

## Section 8: The Alternative Energy Sources

*Storing Solar Energy*

That is the question. Will there ever be a practical way to store the excess solar energy generated during the summer months for use in the winter months? The prospects for doing this with batteries in the near term are dim (pun intended). As pointed out in Figure 6-3b, the energy-storing capacity of even the most advanced battery is feeble in comparison to the energy in an equivalent weight of fossil fuel.

There are always the time-tunnel solutions to the solar energy storage problem: "Let's build a vast solar capability to make cheap hydrogen for our homes and our cars," or "Let's use solar energy to pump water into giant reservoirs to generate hydroelectric power during the winter," or "Let's build better batteries." These may be feasible in the long term, but they don't appear to be near-term solutions for us – our grandchildren or children's grandchildren maybe.

### Wind Energy Generation

What are the limiting factors for wind energy generation? Is it the efficiency of wind machines to generate electricity? Is it their fabrication, installation and upkeep costs? Is it the noise they make plus not looking too good on the landscape (at least in the eyes of some)? Is it the relatively large number of wind machines required to provide the electricity we need? No, these problems are being solved. These are not the inherent limitations. Wind and solar energy have the same basic limitations.

*Wind Energy Limiting Factors*

The limiting factors for wind-generated energy are;

1. Wind machines produce only electricity and electricity represents less than 40 percent of all energy consumption.

## Section 8: The Alternative Energy Sources

2. Wind inconsistency – even the windiest areas will have dead times when backup from conventional energy sources is required. Like solar systems, wind machines are part-time workers; they just show up for work a little more often.

3. Wind machines, as with all other electricity generating systems, have no viable method for storing the energy produced duringt peak operating times.

### Biofuel Energy Production

What are the real problems with the large-scale conversion to biofuels to provide for the world's transportation needs? Is it the cost of biofuels compared to petroleum? Is it the public's acceptance of biofuels? Is it the foot-dragging on the part of the big oil companies? Is it the fact that biofuels like ethanol require relatively large amounts of energy to grow and produce?

No, biofuel costs are becoming more competitive with gasoline. Governments and large corporations are making a concerted effort to reduce costs and increase production. The public is now more willing to accept biofuels as the cost of oil keeps rising.

*Energy to Make More Energy*

The arguments over how much energy is required to produce ethanol continue even though the USDA has issued an official positive energy output statement. Experts on both sides of the fence are still dueling over this as we drive to Starbucks for our morning coffee. For every study that proves ethanol is efficient to produce (more energy out than needed to produce), someone else comes up with a study that proves the opposite (Pimentel, Ref. 6).

The amount of energy required to grow and produce ethanol (or any of the other biofuels) will not be the limiting factor for their future use as fuel for the world's increasing transportation needs. What are the limiting factors?

## Section 8: The Alternative Energy Sources

*Biofuel's Inherent Limitations*

The limiting factors for ethanol production are simply the world's *land, water and nutrient resources.*

*Why Can't We Grow All We Want?*

What do you mean; there is plenty of land and sunshine, like in our own Southwest, to grow all the biofuel that we would ever need. Maybe, but how about the water that will be required? The Rio Grande River is running nearly bone dry by the time it reaches the Mexican border. How about fertilizer and nutrients which are currently derived mostly from fossil fuels?

Producing enough biofuel to supply today's transportation needs would be a prodigious job. But how about thirty to forty years downstream when the world will really need a replacement for gasoline? The world population will have increased to over eight billion people. The demand for gasoline will have increased to a trillion gallons per year. You know all those bicycles in China? Most of them won't be around then.

*Ethanol, an Under-Achiever?*

As you may have been reading, ethanol production from corn may be causing more environmental problems than it is solving (Refs. 1, 8 and 12). Another issue with biofuels, like ethanol, is that a gallon has only about 60 percent of the energy of a gallon of gasoline. In other words, almost 40 percent more ethanol will have to be produced to meet energy demands compared to the equivalent amount of gasoline.

*How Much Land Would It Take?*

Everyone who considers the future use of ethanol or other biofuels usually presents data on crop yields and the conversion factors for the production of biofuels. Estimates are then presented on how much land it would take to produce a significant portion of the world's motor-vehicle energy.

## Section 8: The Alternative Energy Sources

No matter what data or what assumptions are used, the answer always comes out to be a very large percentage of the world's arable land. Here is my estimate, not that we need yet another one.

*My Estimate, Agricultural Land for Ethanol*

First let's look at how much ethanol can be produced per unit land area assuming that *no extra energy* is required to produce it. The next step is to estimate how much of the world's arable land would be required to supply future transportation needs, say, thirty years from now.

The current world demand for gasoline is about 600 billion gallons per year. By 2038 the demand could be 900 billion gallons per year. In terms of energy this would be equivalent to 112 QBTU. Using optimistic values for converting corn to ethanol:

*130 bushels of corn per acre times 3 gallons of ethanol per bushel equals 390 gallons per acre. At $7.6 \times 10^4$ BTU per gallon, this equates to $3 \times 10^7$ BTU of ethanol energy per acre.*

Using your pocket calculator gives nearly four billion acres required to supply the expected world demand for motor vehicle energy in the year 2038. This represents over half of the world's six billion acres of arable land [(1)] and assumes *that no extra energy is required for growing and production.* If it is then assumed that additional biofuel will be required to grow and produce the biofuel needed for transportation, the arable land estimate goes off the scale.

---

*(1) Estimates range from four to eight billion acres depending on the definition of "arable."*

## Section 8: The Alternative Energy Sources

Using sugar beets or some other more efficient source material can increase the energy per acreage yields by two times or more. Also the efficiency of biofuel production processes can be expected to be improved by 2038. But considering the world's population approaching nine billion by 2038, it appears that it will come down to the choice of fuel for our cars or food for us.

*Cellulosic Ethanol*

Biofuel researchers say they will find source materials that have higher yields and can be grown on marginal or *unused* farmland. They say agricultural waste products, wood chips and other cellulosic materials will be used in place of corn or soy beans to produce biofuels. Enzymes, bacteria or termites will be employed to produce large scale quantities at low cost. This seems plausible, but when?

There is promising information from recent USDA studies on the use of switchgrass (a cellulosic material) to produce ethanol (Schmer, Ref. 15). These studies indicate that switchgrass can successfully be grown on marginal farmland, and that the energy to produce ethanol from it may be significantly less than from traditional materials such corn or soy beans. Although cellulosic ethanol has been produced in limited amounts for many years, producing large quantities at low cost has been elusive.

The issues for cellulosic ethanol are: how much farmland, marginal, *unused* or otherwise, will be available when the global population reaches nine billion?; how much fossil-fuel energy would be required for growing and production?; what are the long-term environmental effects?; will cellulosic ethanol help us in the near term or is it something for our children's grandchildren?

# Section 8: The Alternative Energy Sources

## Hydrogen and the Fuel Cell

Hydrogen is not a source of energy in itself. A prime source of energy is required to produce it. Natural gas and coal are currently used to produce most of the world's hydrogen using the steam reformation process. Very little hydrogen is generated using electrolysis, which is not a commercially cost-effective process.

The conversion of prime source energy to hydrogen is accompanied by cumulative energy losses: natural gas or coal to hydrogen gas, hydrogen gas to a compressed or liquefied state with accompanying storage tank losses and then the conversion into electricity for its final use. In Section 6 estimates were made for the conversion losses related to hydrogen production as used for fuel-cell vehicles.

### *The Limiting Factors for Hydrogen Energy*

The limiting factors for hydrogen energy are not the technologies and costs involved. The issues of production, storage, distribution and safety and cost can be solved. The limiting factor is where do we get the prime source energy to make all of the hydrogen we will need? In the near term it will have to come from natural gas or coal conversions. For the far term we can use our "time tunnel escape hatch" and say that the energy to make hydrogen will come from all of those new solar, wind and nuclear plants that are going to be built . . . . sometime in the future.

## Section 8: The Alternative Energy Sources

### Geothermal Energy Generation

*The Limiting Factors for Geothermal Energy*

The limiting factors for geothermal energy are site availability and extraction capability. Extraction technology is expected improve markedly in the upcoming decades. But because geothermal energy is indigenous in nature – mostly suited for use at the source location – large scale use in distant urban locations is limited. Geothermal energy can be used as a direct source of heat in addition to providing electricity. Also geothermal energy, unlike solar and wind, is a near full-time worker. The estimate for potential *extractable* geothermal energy is astronomical – 200 to 2000 ZJ! – more than we could ever use for thousands of years. This estimate, however, has little meaning for us at this end of the time tunnel.

### Summary of Section 8

Although the alternative energy sources are growing rapidly in their own right, they still face stiff competition from the old fossils. The alternatives are predicted (by the EIA) to remain a relatively small percentage of global energy production, at least for the near term (now through 2030).

The limiting factors for solar and wind energy are not technological. Because they only produce electricity and are part-time workers, they can never provide but a relatively small portion of total global energy needs.

The limiting factors for biofuel energy production, at least for the near term, are the world's arable land and water resources. Cellulosic-based biofuels may someday overcome these limitations, but for now they appear to be in the far-term category.

# Section 9

# How to Use This Handbook

When reading about energy-saving proposals, the savings are usually given for a specific fuel or commodity, e.g., gallons of gasoline saved, barrels of oil saved or kilowatts of electricity saved. So, we save 60 million or 60 billion gallons of gasoline by doing such and such. What does this mean? Rarely are these savings translated to the global perspective. In this section we will do just that – we will turn the telescope around to see what's actually happening at the global level with energy savings.

## Some "What-If" Energy-Saving Examples

In this section several of the common energy saving proposals will be examined to determine the effect on global energy consumption. "What if everyone did this . . ." is the name of the game we will play. Keep in mind that while these may be hypothetical examples, they do make a point about the difficulty of saving energy at the global level. Also note that the examples are based on the *Magic Wand Approach.* With a wave of the wand everyone in the world complies with the energy-savings proposal.

## Example 1: Fluorescent Bulbs

As the first example, let's look at using fluorescent instead of incandescent bulbs at home to reduce world energy consumption and to help save the environment at the same time. There will be some energy savings, but the question is how significant would they be? By using the figures and tables from Section 3 (Where Does All the Energy Go?), along with some simple arithmetic, we can estimate the energy savings resulting from switching to fluorescent bulbs. The energy consumption values used in this and the other examples to follow are taken from Table 3-2 and Figure 3-7.

*Data and Assumptions*

*Energy consumption per bulb is about 50 % less than for incandescent bulbs.*

*World energy consumption for residential lighting is about 4.0 QBTU-EUC.*

## Section 9: How to Use This Handbook

*World prime source energy for residential lighting is about three times EUC or 12 QBTU prime source energy – remember the energy-flow chain?*
*World prime energy production is 490 QBTU.*

<u>*Global Energy Savings*</u>

*The energy saved would be 6 QBTU yearly (50% of 12 QBTU).*
*The percent energy savings would be 1.2% (6 out of 490 QBTU).*

> *If every household in the world switched to fluorescent bulbs, global prime source energy use would be reduced by a little over one percent.*

You may have read using fluorescent bulbs could remove the equivalent of so many cars from our highways, let's say 500,000 for example. Sounds good, but what would this accomplish on the global level? Based on the 800 million or more motor vehicles in the world, this translates to a 0.06 percent reduction in equivalent gasoline consumption. This in turn would be a 0.01 percent savings in fossil fuel use, gasoline for motor vehicles being only 14 percent of all fossil-fuel consumption. In case you are wondering, we use fluorescent bulbs in our house.

**Example 2: Clothes Dryers**

What if everyone in the world used clotheslines instead of clothes dryers?

<u>*Data and Assumptions*</u>

*Assume all clothes dryers are electric.*
*World energy used by residential clothes dryers is about 1.3 QBTU-EUC or 4 QBTU prime source energy.*
*World total energy production is 490 QBTU.*

<u>*Global Energy Savings*</u>

*The energy savings in PSE would be less than 1.0% (4 out of 490 QBTU).*

## Section 9: How to Use This Handbook

*If everyone in the world would stop using their clothes dryers, about one percent of global prime source energy would be saved.*

### Example 3: Residential Electricity

What if everyone in the world reduced their household electricity use by ten percent? We are all continuously reminded to save electricity. Turn off the lights when not needed; replace old appliances with energy-efficient ones; turn down the heater at night; set the air conditioner to a higher temperature; don't use so many electronic gadgets; don't wash clothes so often; don't use the electric clothes dryer, etc.

Well, what if everyone – yes everyone in the world – were able to cut their household electricity use by ten percent? That should surely result in significant savings of global energy. Let's see.

*Data and Assumptions*

*World residential electricity consumption is 18 QBTU-EUC.*
*World residential electricity prime source energy is 54 QBTU.*
*(3 times 18 QBTU-EUC = 54 QBTU).*
*World prime energy production is 490 QBTU.*
*World fossil-fuel production is 429 QBTU.*

*Global Energy Savings*

*The energy savings would be 5.4 QBTU (10% of 54 QBTU).*
*The savings in PSE would be 1.1 percent (5.4 out of 490 QBTU).*
*The fossil-fuel savings would be 1.25 percent (5.4 out of 429 QBTU).*

*If every household in the world reduced its electricity use by ten percent, the savings in global prime source energy would be 1.25 percent.*

## Section 9: How to Use This Handbook

You may say, "True, that's not a large savings, but at least it's a start – every little bit counts." I would then remind you that overall global electricity use is growing at 2.2 percent yearly, and that this saving would be washed away in a year or so. What then, ask everyone to save another ten percent?

### Example 4: Reduced Driving

What if all motor vehicle owners reduced their driving by ten percent? Considering that motor vehicles are one of the major users of energy, this idea should be a tremendous saver of oil and fossil fuel, and also a reducer of greenhouse-gas emissions.

*Data and Assumptions*

*World gasoline consumption is 75 QBTU.*

*World gasoline consumption for motor vehicles is 60 QBTU (not all gasoline is used for motor vehicles).*

*World oil consumption is 185 QBTU.*

*World fossil-fuel consumption is 429 QBTU.*

*Global Energy Savings*

*The gasoline energy saved would be 6.0 QBTU (10% of 60 QBTU).*

*The percent of gasoline saved would be 8% (6 out of 75 QBTU).*

*The percent of oil saved would be 3.2% (6 out of 185 QBTU).*

*The percent of fossil-fuel saved would be 1.4 % (6.0 out of 429 QBTU).*

*If all motor vehicle owners reduced their driving by ten percent, global fossil-fuel consumption would be reduced by 1.4 percent.*

## Section 9: How to Use This Handbook

### Example 5: Hybrid Vehicles

What if everyone in the world switched to a high-mileage hybrid vehicle?

<u>*Data and Assumptions – The Magic Wand Approach*</u>

*Hybrids can get twice the mileage of today's average car (44 vs. 22 mpg).*

*World gasoline consumption is 75 QBTU.*

*World gasoline consumption for motor vehicles is 60 QBTU.*

*World oil consumption is 185 QBTU.*

*World fossil-fuel consumption is 429 QBTU.*

<u>*Global Energy Savings*</u>

*The savings in gasoline energy would be 30 QBTU (50% of 60 QBTU).*

*The percent of gasoline saved would be 40% (30 out of 75 QBTU).*

*The percent of oil saved would be 16% (30 out of 185 QBTU).*

*The percent of fossil-fuel saved would be 7% (30 out of 429 QBTU).*

Figure 9-1 (next page) presents another way of looking at the benefits of improved fuel efficiency. This figure is a revised version of Figure 6-13 from Section 6. The original energy entries remain, but have been crossed through, so that easy *before* and *after* comparisons can be made. This makes it easier to visualize what happens on the global level when motor vehicle fuel-efficiency is doubled. The savings in gasoline, oil and fossil fuel are summarized at the bottom of the figure. The figure reminds us that motor vehicles use only about a third of all petroleum and that petroleum consumption in the non-motor vehicle areas will continue to grow regardless of improvements in fuel-efficiency.

This figure clearly exemplifies the meaning of the phrase used earlier, "We have been looking at energy through the wrong end of the telescope."

## Section 9: How to Use This Handbook

By focusing only on the gasoline saved, we miss that motor vehicles use only 32 percent of all petroleum and that petroleum is only 42 percent of all fossil fuel consumption. The flow down from fossil fuel to petroleum to motor-vehicle gasoline results in only a 7 percent savings of fossil fuel even though fuel-efficiency has been doubled for *every motor vehicle in the world.*

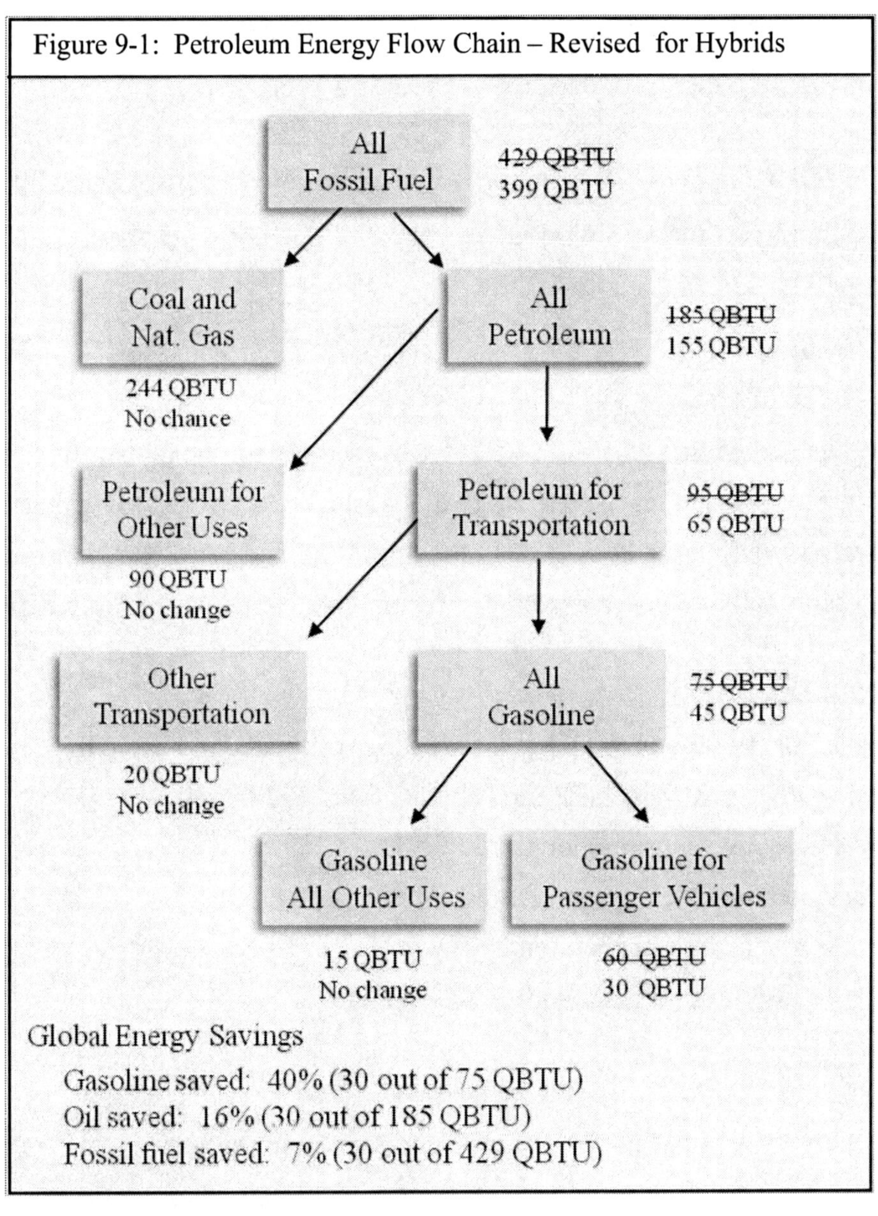

Figure 9-1: Petroleum Energy Flow Chain – Revised for Hybrids

## Section 9: How to Use This Handbook

As pointed out earlier, this estimate is based on the *Magic Wand* approach – all vehicles become energy-efficient with the wave of the wand. Two critical factors have not been considered:

1. The estimate does not account for the increase in motor vehicles expected during the time it would take for the world's fleet to attain 44 mpg average.
2. Oil consumption would continue to grow independently in the "Petroleum for Other Use" category while this was occurring.

The following *Real Life* example is given to show what might actually happen during the time it would take for the world's motor vehicle fleet to reach 44 mpg.

*Data and Assumptions – Real-Life Approach*

It is estimated that it would take 20 years or more for the global fleet of motor vehicles to reach an average of 44 mpg, even with a strong push from world governments. During this time fossil-fuel and petroleum consumption for non-motor vehicle use would still continue to rise (see EIA predictions).

*The 2008 Situation*

The number of motor vehicles worldwide is estimated to be 800 million.
The average vehicle gets 22 mpg and is driven 13,000 miles per year.
World gasoline consumption is 75 QBTU.
World motor vehicle gasoline consumption is 60 QBTU.
World petroleum consumption is 185 QBTU.
World fossil-fuel consumption is 429 QBTU.

*The Scenario for 2033*

The number of world motor vehicles increases to 1.6 billion.
The average vehicle gets 44 mpg and is driven 15,000 miles per year.

## Section 9: How to Use This Handbook

*The Gasoline Savings – Realistic Example*

Based on these assumptions, Figure 9-2 shows how gasoline, oil and fossil-fuel consumption would be affected over the next 25 years as fuel economy was being improved. The "If 22 mpg" assumes average mileage stays at 22 mpg. The "If 44 mpg" assumes the world fleet attains an average mileage of 44 mpg by 2033.

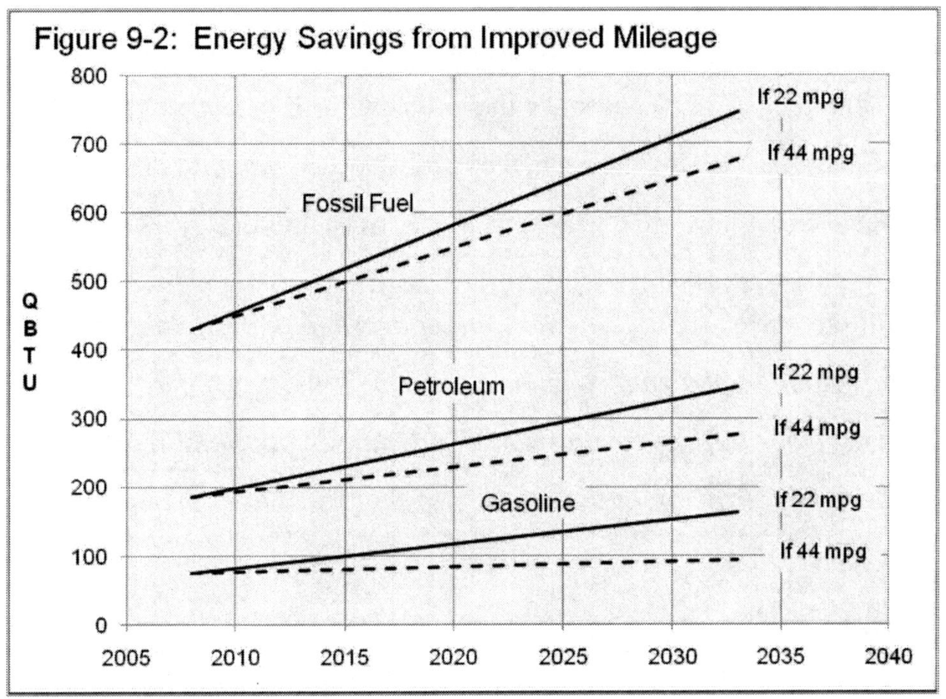


<u>*Summary of Global Energy Savings from Figure 9-1*</u>

*1. Gasoline consumption would be significantly slowed down, but not reduced; it would be 25 percent higher in 2033 than today.*

*2. Petroleum consumption would continue to rise; in 2033 it would be 50 percent higher than today.*

*3. Fossil-fuel consumption would continue to rise; in 2033 it would be 60 percent higher than today.*

## Section 9: How to Use This Handbook

A hypothetical case you say? Considering that fuel economy has almost doubled since the early 1970s, we should be using half as much gasoline as we were using then, correct? No, we are now using nearly twice as much gasoline and petroleum as we were in 1970. This is because the number of motor vehicles has more than quadrupled since 1970 and the "other uses" of petroleum have continued to grow unabated.

The real-life case presented here assumes motor vehicles will only double (to 1.6 billion) by 2033. If autos like the forthcoming Tata *Nano* from India are successful (at less than $2,500 each, they probably will be), the number of motor vehicles in the world could be well over two billion.

> *If all motor vehicles were to attain an average fuel economy of 44 mpg; global petroleum consumption would still be on the increase due to the growing world automobile population and the growth of petroleum in the other (non-motor vehicle) areas.*

**Example 6: Electric Vehicles**

What if there was a major world-wide switch to electric vehicles? You have probably heard the simple answer to this energy-saving proposal. It goes something like this: vast amounts of gasoline and fossil fuel would be saved and greenhouse emissions would be drastically reduced.

There have been countless analyses using sophisticated models to estimate the potential fossil-fuel savings of switching to electric vehicles. The following is a simplified estimate using the parameters and assumptions for electric vehicles as presented in Section 6.

## Section 9: How to Use This Handbook

<u>*Data and Assumptions*</u>

1. Motor vehicles account for 500 billion gallons of gasoline or about 60 QBTU in 2007. If half the world switched to electric vehicles, 30 QBTU (of gasoline energy) could be saved. This saved gasoline energy would have to be replaced with some amount of new electrical energy.

2. Since a typical mid-sized electric vehicle is about four times more energy efficient (EUC) than a typical gasoline vehicle of the same approximate size, the 30 QBTU of gasoline saved could be replaced with about 8 QBTU-EUC of electrical energy.

3. This 8 QBTU-EUC, in turn, translates to 24 QBTU of additional prime source energy for electricity (*Energy Flow Chain*, Figure 3-7).

4. The additional energy would be supplied from the existing electrical grid through the increased use of conventional energy sources, mainly coal and natural gas – this at least until the grid's use of nuclear, hydroelectric and alternates energy sources could be expanded to cover electric vehicles.

<u>*Global Energy Savings*</u>

Currently about 133 QBTU of fossil fuel goes into the generation of global electricity. An increase of 24 QBTU (18 percent) in coal and natural gas consumption would be required for the switch to 50 percent electric vehicles.

*The global gasoline savings would be 40 percent (30 out of 75 QBTU).*

*The savings in petroleum would be 16 percent (30 out of 185 QBTU).*

*The fossil-fuel savings would be 1.4 percent (30 less 24 out of 429 QBTU).*

*Revised Petroleum Energy Flow Chain*

The following is a different way of looking at the above example. Figure 6-13, from Section 6, has been modified to show the effects on global energy consumption resulting from a major switch to electric vehicles.

**Section 9: How to Use This Handbook**

Figure 9-3: The Petroleum Energy Flow Chain – Fifty Percent EVs

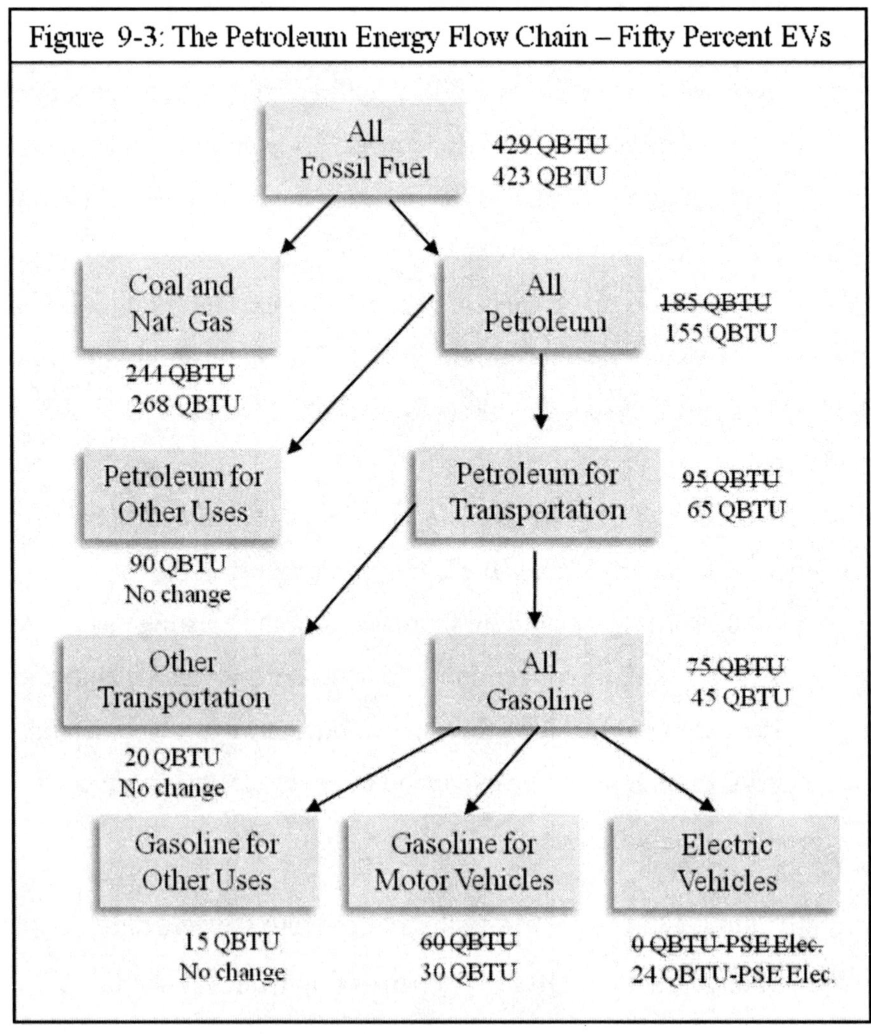


Gasoline saved; 40% (30 out of 75 QBTU)
Oil saved; 16% (30 out of 185 QBTU)
Fossil fuel saved; 1.4% (6 out of 429 QBTU – 30 QBTU saved gasoline plus 24 QBTU new fossil fuel for electricity).

Figure 9-3 shows how gasoline, petroleum and fossil-fuel consumption would be changed due to a major switch (50%) to electric vehicles. The original entries of the *Petroleum Energy Flow Chain (*Figure 6-13) remain, but have been crossed through so that easy *before* and *after* comparisons can be made. The global energy savings are summarized at the bottom of the figure.

## Section 9: How to Use This Handbook

The assumptions for this revised figure are: fifty percent of all motor vehicles switch to electrical power; the additional electrical energy is supplied by the grid using conventional sources. The inevitable increase in overall motor vehicle population and the growth of petroleum in the "other use" areas have not been included.

*Looking at the Revised Petroleum Flow Chain*

The revisions indicated in Figure 9-3 point out that switching to electric vehicles must be viewed from the global energy perspective. Focusing only on how much gasoline could be saved masks the problem of global energy consumption and savings.

The savings in global energy is less than we have been led to believe. From the diagram it is seen that although significant amounts of gasoline would saved, the overall fossil fuel savings would be small, only about 1.4 percent. The savings in petroleum would largely be offset by the increased use of coal and natural gas for electricity generation.

Also the non-motor vehicle uses of petroleum and fossil-fuel would continue to grow independently during the switchover to electric vehicles, thus even further reducing the effectiveness of this energy saving proposal.

**Example 7: Proper Tire Inflation**

For the final example, the energy that could be saved if everyone kept their automobile tires properly inflated will be estimated. This energy saving idea gets a lot of arm-waving attention, but rarely is it accompanied with realistic estimates of the actual energy that could be saved. A few basic assumptions and some simple arithmetic will be needed to make this estimate.

## Section 9: How to Use This Handbook

*Data and Assumption:*

*The number of world motor vehicles is 800 million.*

*Assume 12 percent of above have under-inflated tires (96 million vehicles).*

*Assume proper inflation increases mileage by 12 percent (19 to 21.5 mpg).*

*World gasoline consumption is 615 billion gallons (75 QBTU).*

*World petroleum production is 185 QBTU; fossil fuel is 429 QBTU.*

*Global Energy Savings*

*The gasoline savings would be 7 billion gallons yearly.*

*The gasoline energy savings would be 0.9 QBTU.*

*The gasoline percent savings would be 1.2% (0.9 out of 75 QBTU).*

*The oil percent savings would be 0.5% (0.9 out of 185 QBTU).*

*If all motor-vehicle tires were properly inflated, the percent savings in fossil fuel would be 0.2% (0.9 out of 429 QBTU).*

**Summary of Section 9**

Energy saving proposals, which initially appear so promising, are diluted in effectiveness when viewed from the global perspective.

The “Energy Savings Begin at Home” editorials that we are so familiar with seem not to recognize that households use only 15 percent of all consumed energy.

Improvements in fuel efficiency, though highly commendable, have not been able to keep up with the ever-increasing population of motor vehicles.

Switching to electric vehicles, while saving a significant amount of gasoline, would have a far-less impact on reducing global oil and fossil-fuel consumption.

# Section 10

# The Energy Triad

As we all know, energy consumption, the economy and the environment are closely intertwined. In this section simple relations are given which will help us see how closely the elements of the Energy Triad are tied together.

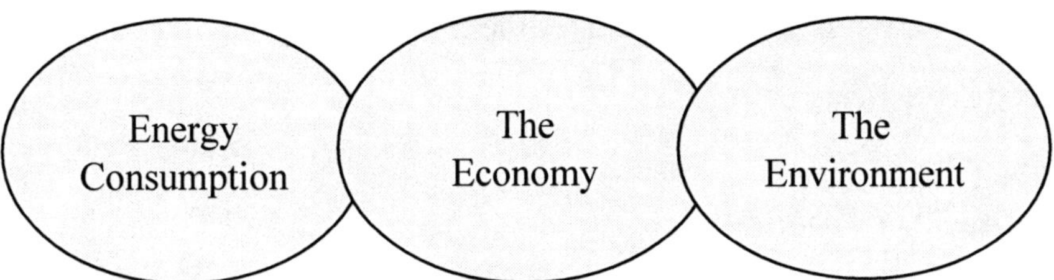


## Global Energy Consumption – Past and Future

As was seen in Section 7 (Looking Back at Energy Savings), global energy use has been on a continual rise, even after the energy crises of the 1970s set off a plethora of conservation, efficiency and alternative energy initiatives.

Figure 10-1: World Prime Source Energy Growth (from EIA Figure 10)

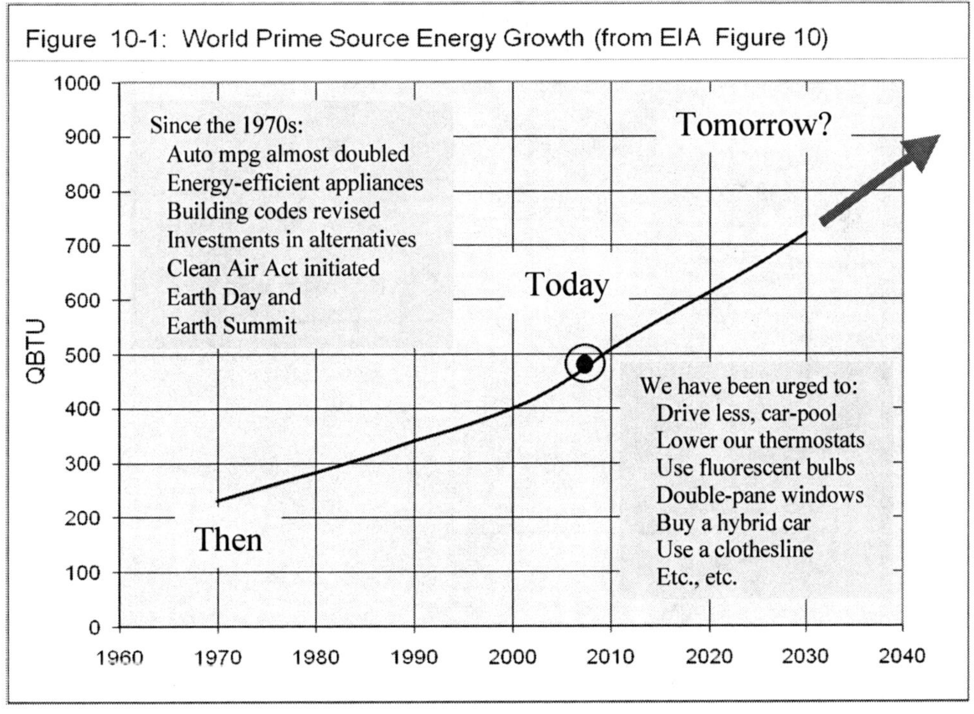

## Section 10: The Energy Triad

This rise in global energy consumption continues despite the many conservation practices and energy-saving technologies introduced since the early 1970s. If predictions are anywhere near correct, the climb will continue for the next several decades. Figure 10-1 shows how global energy has grown since 1970. The figure also summarizes the many energy-savings initiatives introduced since then.

### Energy and "The Economy"

Even if you are not an economist, energy expert or environmental scientist, you still intuitively know what this simple diagram, Figure 10-2, is suggesting

Figure 10-2: The Energy Triad – A Precarious Balance

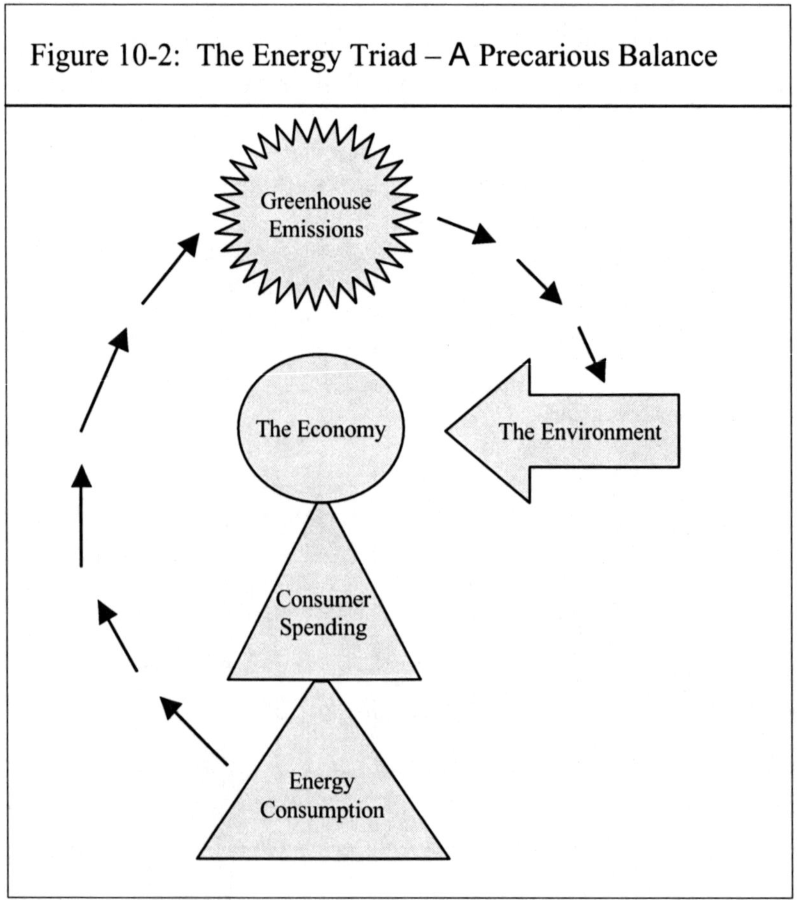

## Section 10: The Energy Triad

The Economy is supported (precariously) by consumer spending, which is fueled by energy consumption (what else?), which in turn results in greenhouse emissions which cause . . . you know what. That's the sequence of events; now let's see how the elements are related numerically.

But first, it is noted that the diagram suggests the way it is, not the way it always has to be. As you know, maintaining economic growth while reducing energy consumption and greenhouse emissions has become the major challenge (and controversy) of our time.

### Fossil-Fuel Growth

First, let's look at fossil-fuel consumption. Figure 10-3 shows the world's fossil-fuel energy production since 1970, with EIA estimates out to the year 2030.

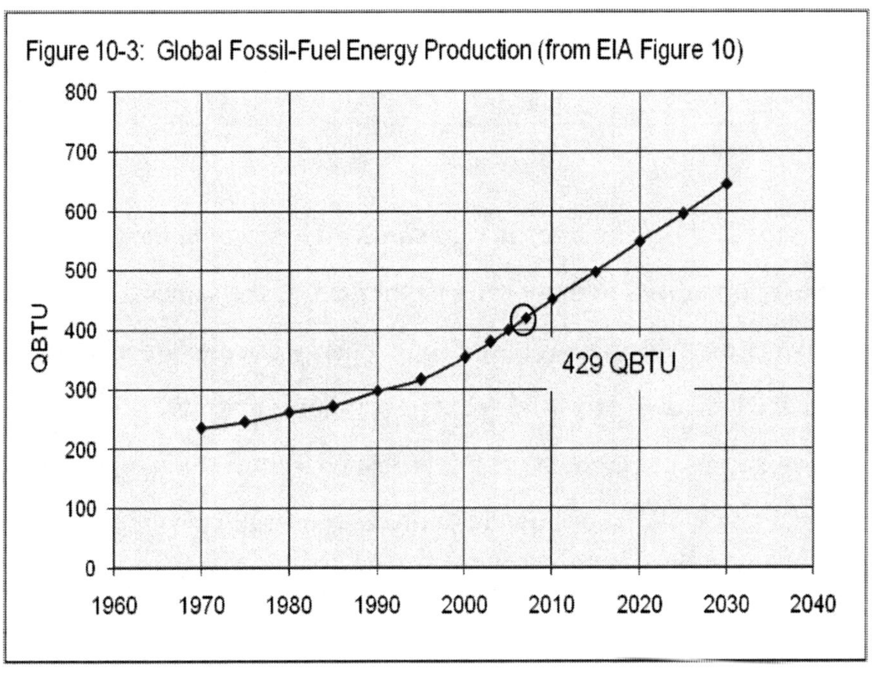

# Section 10: The Energy Triad

## Greenhouse Emissions

Figure 10-4 gives the past and estimated future global greenhouse emissions in terms of billions of tons of $CO_2$.

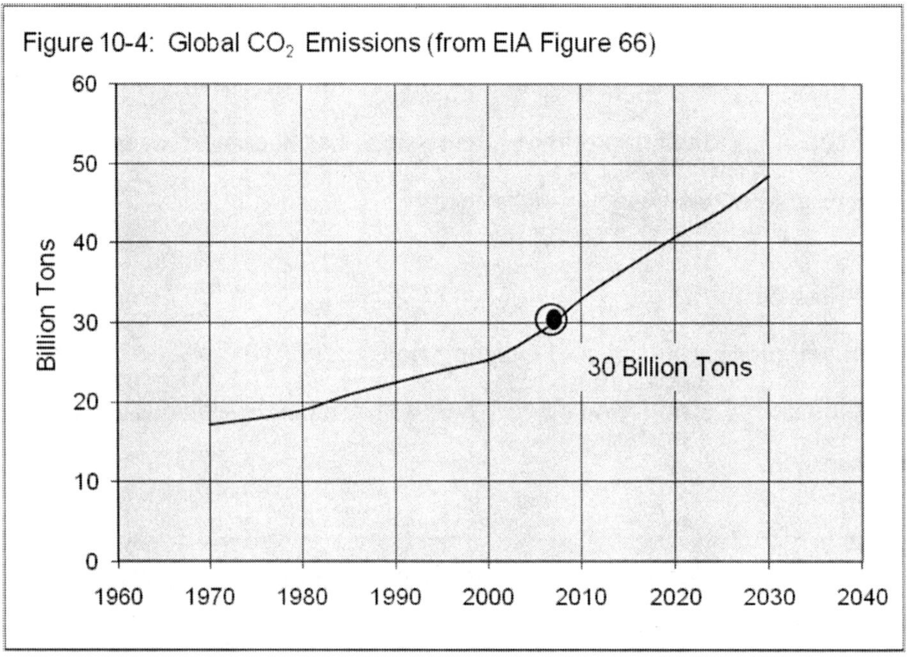


Figure 10-4: Global $CO_2$ Emissions (from EIA Figure 66)

The shape of the two curves is very similar. How would the two curves look if they were normalized to unity and then plotted on the same axes? Doing this should give us an indication of how closely energy use and greenhouse emissions are linked. Figure 10-5 (next page) shows that fossil-fuel consumption and $CO_2$ emissions are more than closely related – they increase in lockstep with each other. Next we will see how these relate to overall consumer spending.

## Section 10: The Energy Triad

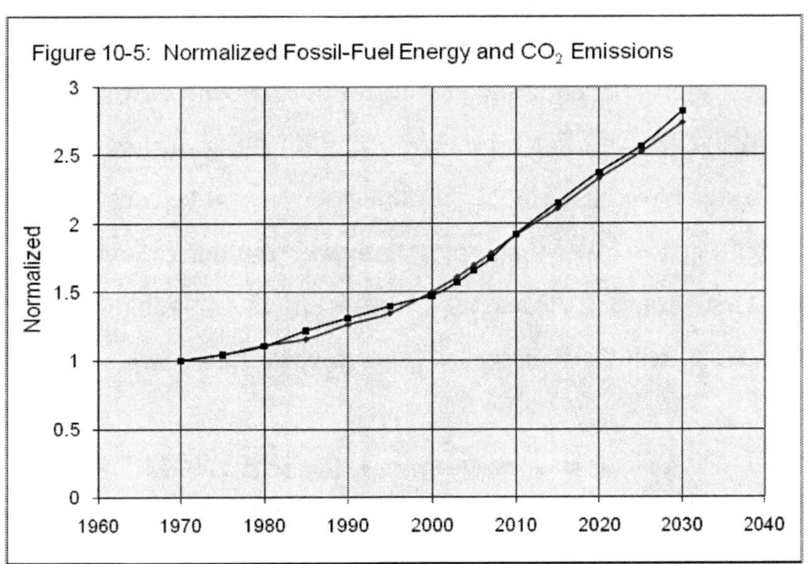


### The Gross World Product, GWP

Figure 10-6 is a plot of consumer spending expressed as Gross World Product or GWP. It was compiled from several sources and may not be considered by some to be the best data for comparing energy consumption and greenhouse emissions with the global economy, but here it will serve to make a simple, if not academic, point.

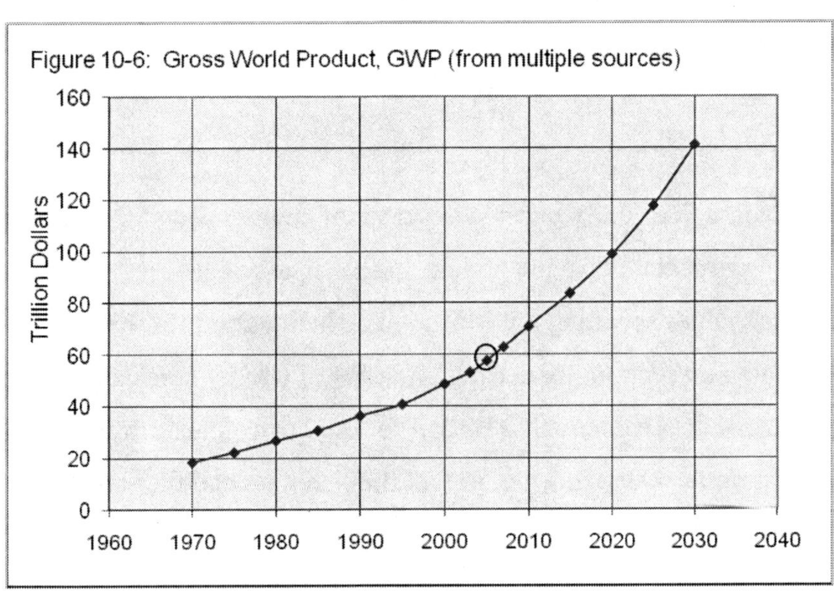

# Section 10: The Energy Triad

## The Energy Triad Plotted

Figure 10-7 shows the combined results of normalizing Gross World Product (GWP) data with fossil-fuel consumption and greenhouse-emissions data. The figure shows that global economic activity (at least as measured by my GWP data) grows faster than energy use and resultant greenhouse emissions. That's a relief! At least it's comforting to know that not all global economic activity results in increased greenhouse emissions.

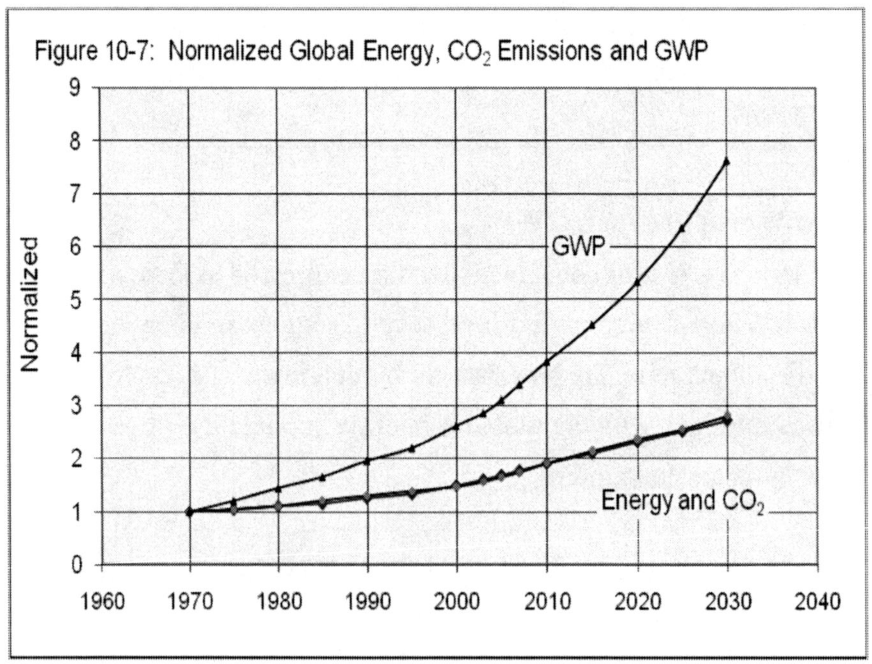

Figure 10-7: Normalized Global Energy, $CO_2$ Emissions and GWP

When reading your daily paper, the financial news is found in one section and the environmental news in another. In the financial section, reports on increased consumer spending are looked at as being good for the economy. In the environmental section, increasing amounts of energy consumption and greenhouse gasses are deemed to be bad for the environment. It is as if the two issues were not related, when in fact they are one and the same.

## Section 10: The Energy Triad

### What's Not the Solution?

The purpose of this handbook is not to suggest solutions to our global energy problems – an army of experts around the world is working on this.

The purpose of this handbook is to provide a concise and usable set of energy information allowing you to understand, and challenge when necessary, what the experts are saying. This handbook may not give you the answers, but hopefully it will help you in asking the right questions.

*Asking the Right Questions*

When you run across a magazine or newspaper article such as "Conservation and Efficiency Can Solve Our Energy Problems" or "100 Ways to Go Green at Home," remind the writers that everything they are recommending has been recommended for the last three decades or more. Also remind them to translate their energy information to something the readers can relate to. An article entitled "Wind Energy Capacity Increases by 5000 Megawatts" will mean nothing to most people unless translated to the global level.

*Is the Solution Energy Conservation?*

From what we have seen, the solution is not simply everyone switching to fluorescent bulbs or using clotheslines to dry their clothes. It is not simply everyone driving hybrid cars. It is not simply turning down our thermostats. It is not simply asking everyone to conserve energy. Each of these approaches has had little effect on global energy consumption (Section 7, Energy Savings). In addition, any gains made by conservation are quickly washed away by the ever-increasing demands for energy throughout the rest of the world.

## Section 10: The Energy Triad

*Is the Solution in the Alternative Energy Sources?*

Advances in technology will solve most of the issues now being addressed in regard to the alternative energy sources – improvements in efficiency, safety, cost and practicality will certainly occur. However, there are inherent limitations for the alternatives that have nothing to do with technology. (Section 8, Alternative Energy Sources). Solar and wind only produce electricity, and they are part-time workers. Biofuels will require large amounts of the world's arable land and water resources to produce fuel for future global transportation needs – this regardless of how efficient growing and production processes may become.

*Is the Solution in Going Green?*

The green economy is with us. From Mom and Pop businesses to the global giants, everyone is going green. From hybrid cars to solar panels to recycled household items, there is profit to be made in going green. The real question is, however, will all this going green have any effect on reducing or even slowing down the demand for global energy? We won't know the answer to this question until we are able look at the 2018 version of Figure 10-1.

*Is the Solution Reduced Consumer Spending?*

And finally, the answer is not simply asking (or expecting) everyone in the world to drastically reduce their consumer spending. Even though the relationships between consumer spending, energy consumption and greenhouse emissions are obvious, this is just not going to happen. You may have noticed lately that when consumer spending does not rise as *high* as expected the whole economy shudders – imagine if consumer spending were to head south.

## Section 10: The Energy Triad

### Another Game to Play

Earlier in this handbook we suggested the "How Much Gozinto" game to be played with your friends. Here is another game for your friends – assuming you have any left after the Gozinto game. At your next party ask your guests what they think about these questions. Offer those with the correct answers a prize – such as a free tank of gasoline.

The game is called "How many of you think if . . . ."

1. If we all drove high-mileage cars we could drastically reduce our dependence on foreign oil and greenhouse emissions?

2. If we all switched to fluorescent bulbs we could significantly reduce fossil-fuel use and greenhouse emissions?

3. If we built enough solar or wind facilities we could solve our energy problems?

4. If we all drove electric vehicles we could eliminate our dependence on petroleum (foreign or otherwise)?

5. If we built enough nuclear facilities we could solve our energy problems?

6. If we all were less wasteful in our household-energy use we could drastically reduce global energy consumption and greenhouse emissions?

It is doubtful that you will have to buy a tank of gasoline for any of your guests; however as a consolation prize you could give them a copy of this handbook.

## Section 10: The Energy Triad

### Closing Statement

I was listening to a radio talk show while trying think of the last pages for this handbook. The guest was an eminent scholar and writer on environmental and energy issues. The topic was the high cost of gasoline and the proposal to drill for more oil in Alaska. The guest pointed out that this was an exercise in futility, the amount of oil thought to be there supplying the U.S. for only a few years and the world for much less. He said the people proposing this couldn't see the elephant in the room. The elephant, of course, was our *addiction* to oil and the ultimate damage this was going to cause to both the environment and to our economy.

The guest's solution: we must conserve energy; we must be more energy efficient; we must drastically reduce the number of gasoline vehicles on our roads and highways; we must switch over to the alternative energy sources.

I tried to imagine how you, having just read this handbook, would respond to this guest's remarks during the listener phone-in section of the program. Here is what I hoped you would say to him.

First you might agree or disagree on the Alaska drilling issue, but then you would challenge his *solutions*. You would point out that conservation, although the right thing to do, will hardly slow down oil consumption, let alone reduce it. The small benefits from conservation that have occurred over the past decades have been washed away by the increasing demand for oil at the global level.

Improved fuel efficiency has not been able to keep up with the ever-increasing number of motor vehicles. You would also point out that only a third of all petroleum is used by passenger vehicles – the other two-thirds being used for industrial and non-transportation purposes. Petroleum and fossil-fuel use in these non-motor vehicle areas would continue to grow despite gains in fuel efficiency.

## Section 10: The Energy Triad

As for the alternative energy sources, you would ask which ones he was referring to. Solar and wind can only produce electricity. Electricity cannot replace the vast amounts of oil used for industrial and commercial purposes. Electricity is not going run the heavy farm equipment or the trucking, airline and shipping industries which are the basis of the global food chain. It will not supply the raw materials for the petrochemical industry. As for transportation, electric vehicles will be limited to a small number of passenger vehicles (at least until some earth-shaking breakthrough occurs in battery technology).

Ethanol and the other biofuels, as he would surely know, are already impacting global food supplies and the environment. Even with improvements in growing and production, they will become more even competitive with the world's arable land and water resources as the demand for food, along with the world's population, continues to grow (refer him to Section 8). If his reply is that biofuels will be grown on arid or *unused* farmland using very little energy, point out that he is using the "time tunnel escape hatch."

You could remind him that energy does not hold still while we try to *fix* it. By the time we think we have it *fixed*, energy use has grown beyond the fix. Point out that he is looking through the *wrong end of the telescope;* he needs to turn it around to see what's really happening at the global level.

And finally you could say that while the others in the room may not see the elephant, he does not see that there is more than one elephant, there is herd of them.

# Section 11

# Reference Data and Figures

## General References


1. "The Ethanol Myth," Consumer Reports, October 2006, p. 15.

2. "Biofuels: Boon or Boondoggle," Borne, J.K., National Geographic, October 2007, p. 38.

3. "Energy Future Beyond Carbon," Scientific American, Special Issue, September 2006.

4.Green Issue – 2nd Annual, Vanity Fair, Special Issue, May 2007.

5. "Global Warming Survival Guide," Time, April 9, 2007.

6. Pimentel D., Patzek T., "Ethanol Production Using Corn, Switchgrass, and Wood; Biodiesel Production Using Soybean and Sunflower," National Resources Research, Vol. 14, No 1, March 2005 (©2005).

7. Raymond, L.R. et al, "Hard Truths," National Petroleum Council, July 2007, Washington, D.C.

8. International Energy Outlook for 2007, Energy Information Administration (EIA), Washington D.C., 2007. (www.eia.doe.gov: home>international> . . ).

9. Annual Energy Review (AER), EIA, Washington, D.C., 2007. (www.eia.doe.gov: home>historicalenergydata>aer> . . .).

10. Energy Sources, Department of Energy, Washington, D.C. (www.energy.gov>energysources>index.htm).

## Section 11: Reference Data and Figures

11. Demand, Energy Information Administration (EIA), Washington, D.C. (eia home>forecasts&analyses).

12. "The Clean Energy Myth," Michael Grunwald, Time, April 7, 2008.

13. International Energy Agency (IEA), Geothermal Annual Report 2006, Paris, France, January 2008.

14. Kreith, F. and West, R.E., "Mechanical Engineering Power 2003," (Transportation Quarterly, Vol. 56, No. 1, Winter 2002).

15. Schmer, M. R., et al, " Net energy of cellulosic ethanol from switchgrass", Proceedings of the National Academy of Sciences (PNAS), vol. 105, no. 2, January 15, 2008.

16. Small, K.A. and Van Dender, K., "Fuel Efficiency and Motor Vehicle Travel: The Declining Rebound Effect", Dept. of Economics, U.C. Irvine, Irvine, CA, April 10, 2006.

# Section 11: Reference Data and Figures

## Reference Figures and Data Reductions

The following pages document the prime references that were used to generate the figures and tables included in this work. In most cases the original reference sources required graphical reductions and interpretations before being incorporated into the handbook. The reasons for this were:

1. To normalize to the common BTU system used throughout the work.
2. To obtain a better definition for between-year values.
3. To extrapolate data to years beyond those given in the prime data.
4. To make it easier for the reader to get the message of the data.

## Data-Reduction Examples

A typical example of graphical interpretation of prime data is shown for EIA Figure 5 (page 11-12). The original EIA figure is a stacked bar chart showing global electricity generation by fuel type. Only three specific years are given: 2003, 2015 and 2030. The author's derived graph is shown along with the original EIA bar chart. From the derived graph, one can easily read the energy values (QBTU) by year for the fuel types used to generate electricity.

Also note that the title on the original EIA figure reads "World Energy Consumption for Electricity by Fuel Type." The figure actually gives the prime source energy (PSE) sent to the electricity generation system, not the consumed electrical energy as the EIA title implies. Unless you know that 189 QBTU PSE is sent to electricity (as you should by now), this figure could be confused with EIA Figures 57 and 58 (page 11-14), which give end-use electricity consumption (63 QBTU EUC) by the sectors.

## Section 11: Reference Data and Figures

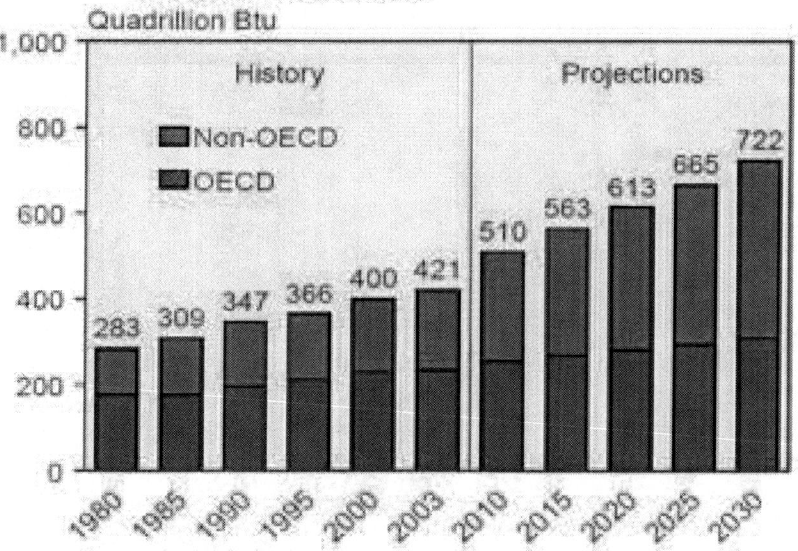


Author's Graphical Data Reduction:

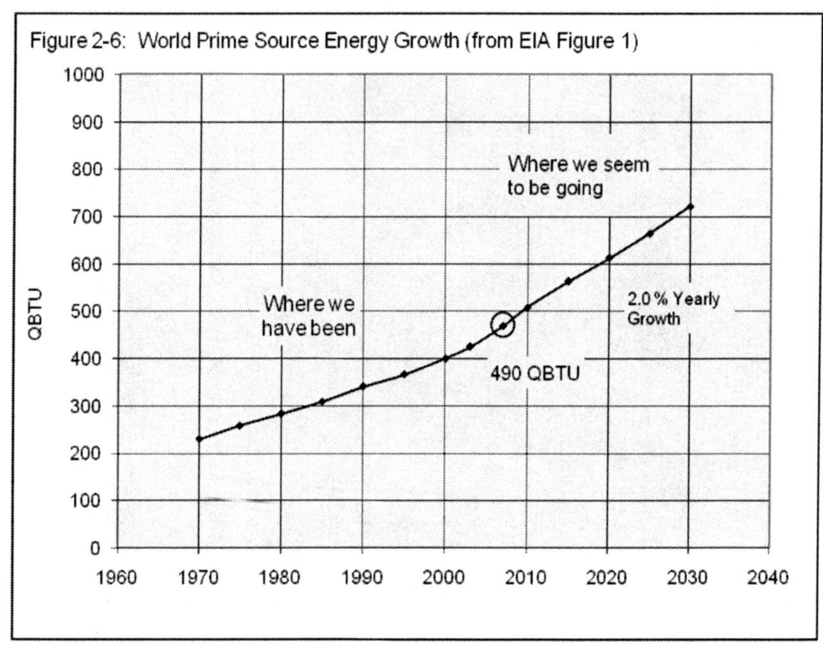

# Section 11: Reference Data and Figures

## EIA Table 11.1

**Table 11.1 World Primary Energy Production by Source, 1970-2004**
(Quadrillion Btu)

| Year | Coal | Natural Gas [1] | Crude Oil [2] | Natural Gas Plant Liquids | Nuclear Electric Power [3] | Hydroelectric Power [3] | Geothermal [3] and Other [4] | Total |
|---|---|---|---|---|---|---|---|---|
| 1970 | 62.96 | 37.09 | 97.09 | 3.61 | 0.90 | 12.15 | 1.59 | 215.39 |
| 1971 | 61.72 | 39.80 | 102.70 | 3.85 | 1.23 | 12.74 | 1.61 | 223.64 |
| 1972 | 63.65 | 42.08 | 108.52 | 4.09 | 1.66 | 13.31 | 1.68 | 234.99 |
| 1973 | 63.87 | 44.44 | 117.88 | 4.23 | 2.15 | 13.52 | 1.73 | 247.83 |
| 1974 | 63.79 | 45.35 | 117.82 | 4.22 | 2.86 | 14.84 | 1.76 | 250.64 |
| 1975 | 66.20 | 45.67 | 113.08 | 4.12 | 3.85 | 15.03 | 1.74 | 249.69 |
| 1976 | 67.32 | 47.62 | 122.92 | 4.24 | 4.52 | 15.08 | 1.97 | 263.67 |
| 1977 | 68.46 | 48.85 | 127.75 | 4.40 | 5.41 | 15.56 | 2.11 | 272.54 |
| 1978 | 69.56 | 50.26 | 128.51 | 4.55 | 6.42 | 16.80 | 2.32 | 278.41 |
| 1979 | 73.83 | 53.93 | 133.87 | 4.87 | 6.69 | 17.69 | 2.48 | 293.36 |
| 1980 | 71.24 | 54.73 | 128.04 | 5.10 | 7.58 | 17.90 | 2.94 | 287.53 |
| 1981 | 71.83 | 55.56 | 120.11 | 5.37 | 8.53 | 18.26 | 3.10 | 282.56 |
| 1982 | 74.25 | 55.49 | 114.45 | 5.35 | 9.51 | 18.71 | 3.27 | 281.05 |
| 1983 | 74.25 | 58.12 | 113.97 | 5.36 | 10.72 | 19.69 | 3.56 | 283.68 |
| 1984 | 78.38 | 61.78 | 116.88 | 5.73 | 12.99 | 20.19 | 3.70 | 299.65 |
| 1985 | 82.20 | 64.22 | 115.37 | 5.83 | 15.30 | 20.42 | 3.78 | 307.13 |
| 1986 | 84.28 | 65.32 | 120.18 | 6.15 | 16.25 | 20.89 | 3.78 | 316.85 |
| 1987 | 86.08 | 68.46 | 121.07 | 6.35 | 17.64 | 20.90 | 3.79 | 324.33 |
| 1988 | 87.94 | 71.80 | 125.84 | 6.65 | 19.23 | 21.48 | 3.96 | 336.90 |
| 1989 | 89.43 | 74.24 | 127.83 | 6.69 | 19.74 | 21.53 | 4.34 | 343.80 |
| 1990 | 90.93 | 75.87 | 129.35 | 6.87 | 20.36 | 22.35 | 3.93 | 349.66 |
| 1991 | 86.29 | 76.69 | 128.73 | 7.12 | 21.18 | 22.83 | 4.03 | 346.86 |
| 1992 | 86.05 | 76.90 | 128.93 | 7.36 | 21.28 | 22.71 | 4.29 | 347.53 |
| 1993 | 84.28 | 78.41 | 128.72 | 7.86 | 22.01 | 23.94 | 4.31 | 349.32 |
| 1994 | 86.27 | 79.16 | 130.56 | 8.27 | 22.41 | 24.15 | 4.49 | 355.32 |
| 1995 | 88.87 | 80.24 | 133.32 | 8.55 | 23.26 | 25.34 | 4.64 | 364.23 |
| 1996 | 88.92 | 83.99 | 136.61 | 8.76 | 24.11 | 25.79 | 4.81 | 372.96 |
| 1997 | 92.15 | 84.29 | 140.52 | 8.94 | 23.88 | 26.07 | 4.91 | 380.75 |
| 1998 | 90.86 | 85.85 | 143.14 | 9.17 | 24.32 | 26.06 | 4.89 | 384.39 |
| 1999 | 90.43 | 87.89 | 140.79 | 9.47 | 25.09 | 26.56 | 5.09 | 385.32 |
| 2000 | 91.36 | 91.34 | 146.55 | 9.87 | 25.66 | 27.01 | 5.35 | 397.13 |
| 2001 | 96.89 | 93.74 | 145.32 | 10.32 | 26.39 | 25.39 | 5.25 | 404.30 |
| 2002 | 97.05 | 96.72 | 143.11 | 10.53 | 26.68 | 26.44 | 5.58 | 406.12 |
| 2003 | 104.81 | 98.93 | 147.97 | 11.02 | 26.45 | 26.83 | 5.90 | 421.71 |
| 2004P | 113.30 | 102.19 | 154.79 | 11.48 | 27.47 | 27.53 | 6.33 | 443.10 |

[1] Dry production.
[2] Includes lease condensate.
[3] Net generation, i.e., gross generation less plant use.
[4] Includes net electricity generation from wood, waste, solar, and wind. Data for the United States also include other renewable energy.
P=Preliminary.
Notes: • Data in this table do not include recent updates for the United States (see Table 1.2) or for other countries (see the Energy Information Administration's "International Energy Annual 2005"). • See Note 1, "World Primary Energy Production," at end of section. • Totals may not equal sum of components due to independent rounding.
Web Page: For related information, see http://www.eia.doe.gov/international.
Sources: • 1970-1979—Energy Information Administration (EIA), International Energy Database. • 1980 forward—EIA, "International Energy Annual 2004" (May-July 2006), Tables F1-F8.

## Author's Graphical Data Reduction:

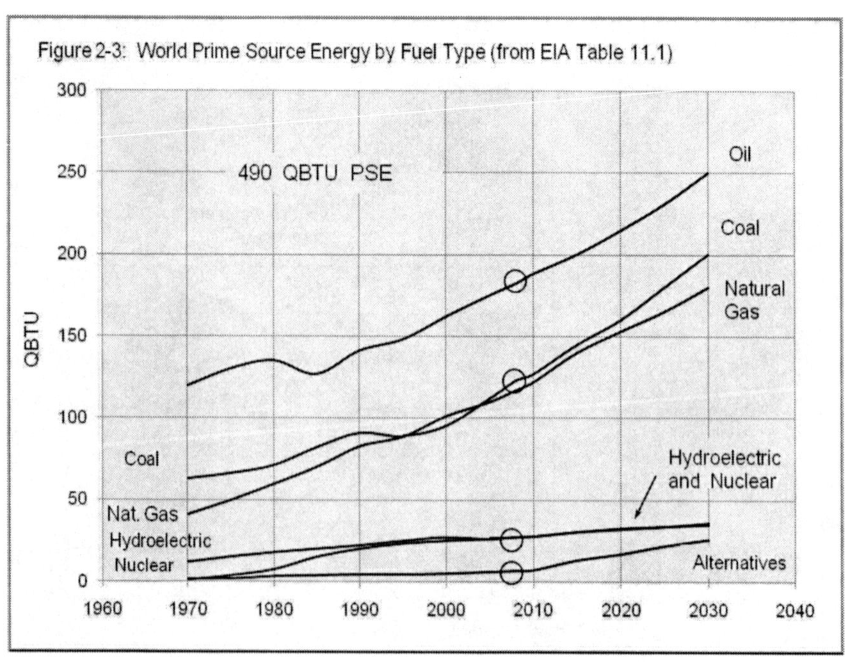

## Section 11: Reference Data and Figures

Figure 4. World Marketed Energy Use by Fuel Type, 1980-2030

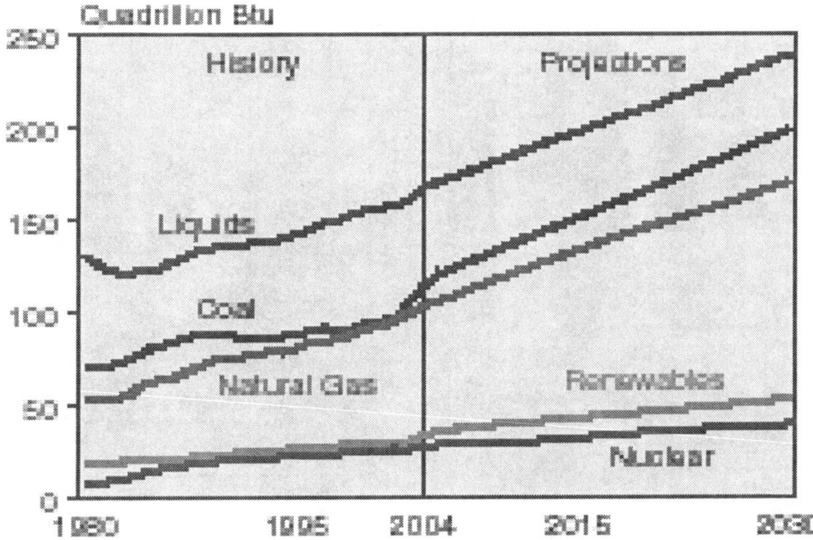


Sources: History: Energy Information Administration (EIA), *International Energy Annual 2004* (May-July 2006), web site www.eia.doe.gov/iea. Projections: EIA, System for the Analysis of Global Energy Markets (2007).

Author's Graphical Data Reduction:

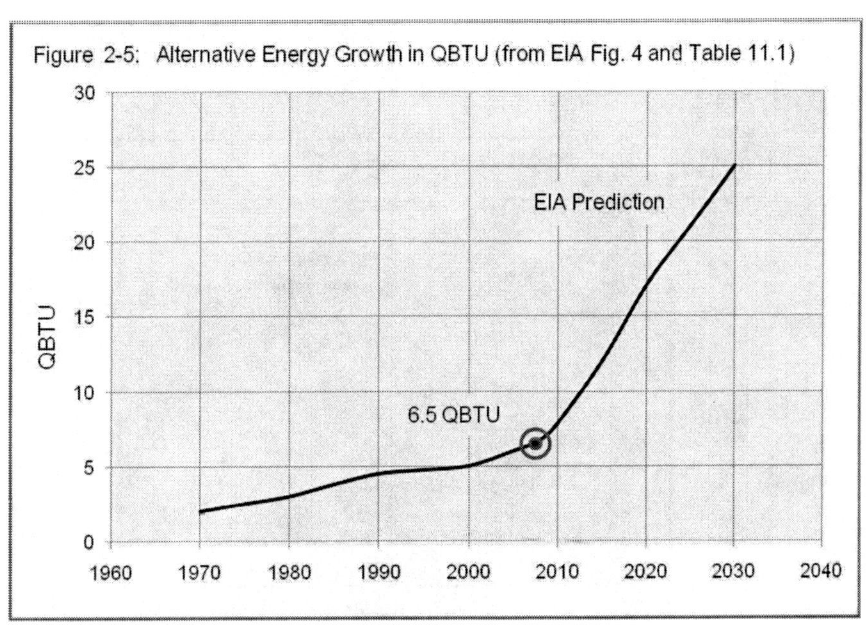


Figure 2-5: Alternative Energy Growth in QBTU (from EIA Fig. 4 and Table 11.1)

# Section 11: Reference Data and Figures

**Figure 19. OECD and Non-OECD Residential Sector Delivered Energy Consumption, 2003-2030**

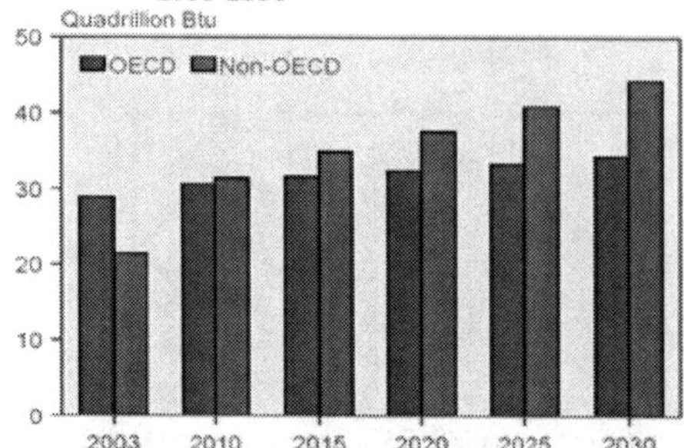


Sources: **2003:** Derived from Energy Information Administration (EIA), *International Energy Annual 2003* (May-July 2005), web site www.eia.doe.gov/iea/. **Projections:** EIA, System for the Analysis of Global Energy Markets (2006).

**Figure 25. OECD and Non-OECD Transportation Sector Delivered Energy Consumption, 2003-2030**

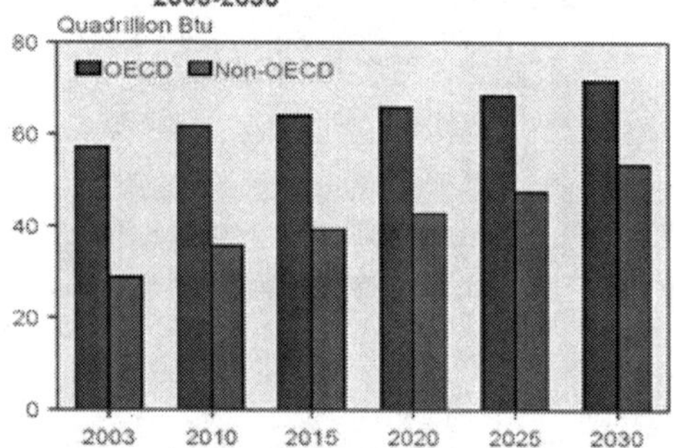


Sources: **2003:** Derived from Energy Information Administration (EIA), *International Energy Annual 2003* (May-July 2005), web site www.eia.doe.gov/iea/. **Projections:** EIA, System for the Analysis of Global Energy Markets (2006).

**Figure 23. OECD and Non-OECD Industrial Sector Delivered Energy Consumption, 2003-2030**

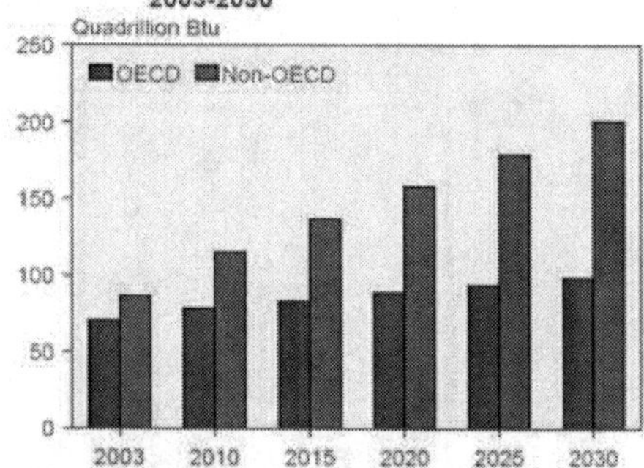


Sources: **2003:** Derived from Energy Information Administration (EIA), *International Energy Annual 2003* (May-July 2005), web site www.eia.doe.gov/iea/. **Projections:** EIA, System for the Analysis of Global Energy Markets (2006).

**Figure 21. OECD and Non-OECD Commercial Sector Delivered Energy Consumption, 2003-2030**

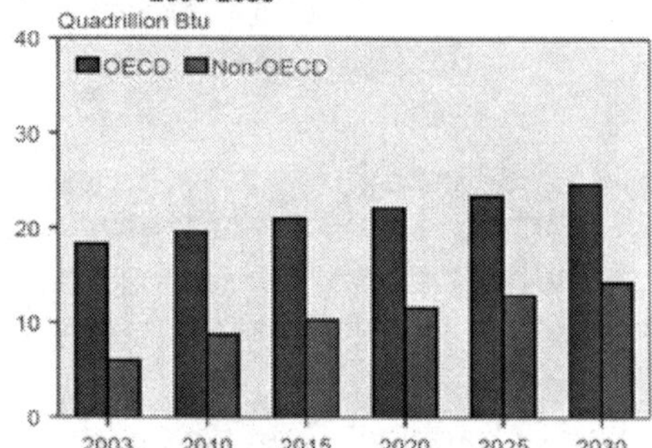


Sources: **2003:** Derived from Energy Information Administration (EIA), *International Energy Annual 2003* (May-July 2005), web site www.eia.doe.gov/iea/. **Projections:** EIA, System for the Analysis of Global Energy Markets (2006).

## Author's Graphical Data Reduction:

Figure 3-4: End-Use Consumption by the Sectors
(from EIA Figures 19, 21, 23, 25)

QBTU

183 QBTU (50%)

The total is 364 QBTU
(losses are not included)

97 QBTU (27%)

55 QBTU (15%)

29 QBTU (8%)

Industrial Transportation Residential Commercial

# Section 11: Reference Data and Figures

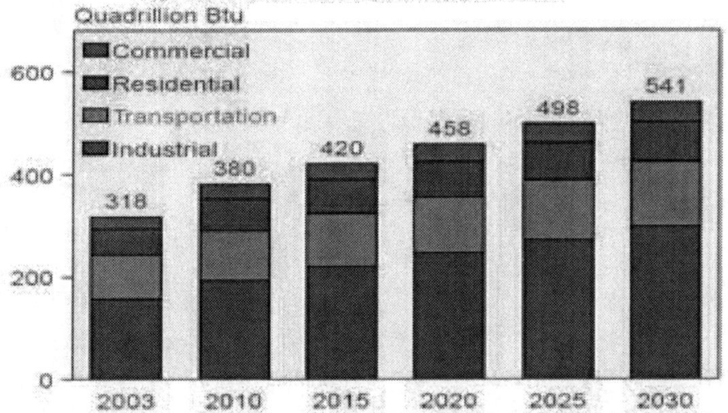


Author's Graphical Data Reduction:

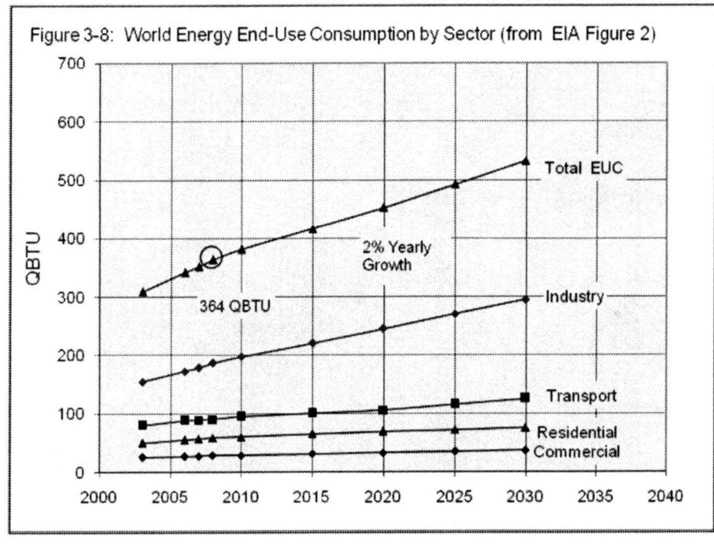


| Table 3-3 : World End-Use Consumption by Year (from EIA Figure 2) | | | | | |
|---|---|---|---|---|---|
| Year | Industry | Transport | Residential | Commer. | Total |
| 2003 | 154 | 80 | 50 | 25 | 309 |
| 2006 | 172 | 88 | 55 | 26.5 | 342 |
| 2007 | 178 | 89 | 57 | 27.3 | 351 |
| **2008** | **186.4** | **90.4** | **58.4** | **28.4** | **364** |
| 2010 | 197 | 95 | 60 | 29 | 381 |
| 2015 | 220 | 100 | 65 | 31 | 416 |
| 2020 | 245 | 105 | 69 | 33 | 452 |
| 2025 | 270 | 115 | 72 | 35 | 492 |
| 2030 | 295 | 125 | 75 | 37 | 532 |

# Section 11: Reference Data and Figures

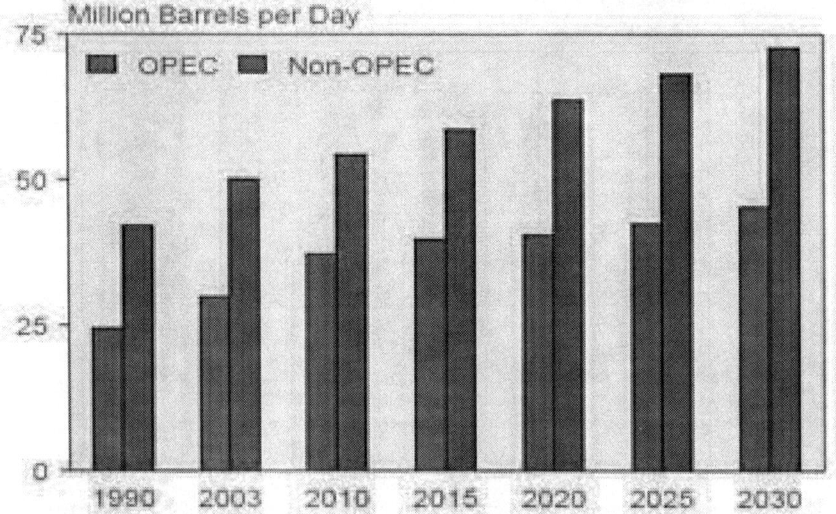


Author's Graphical Data Reduction:

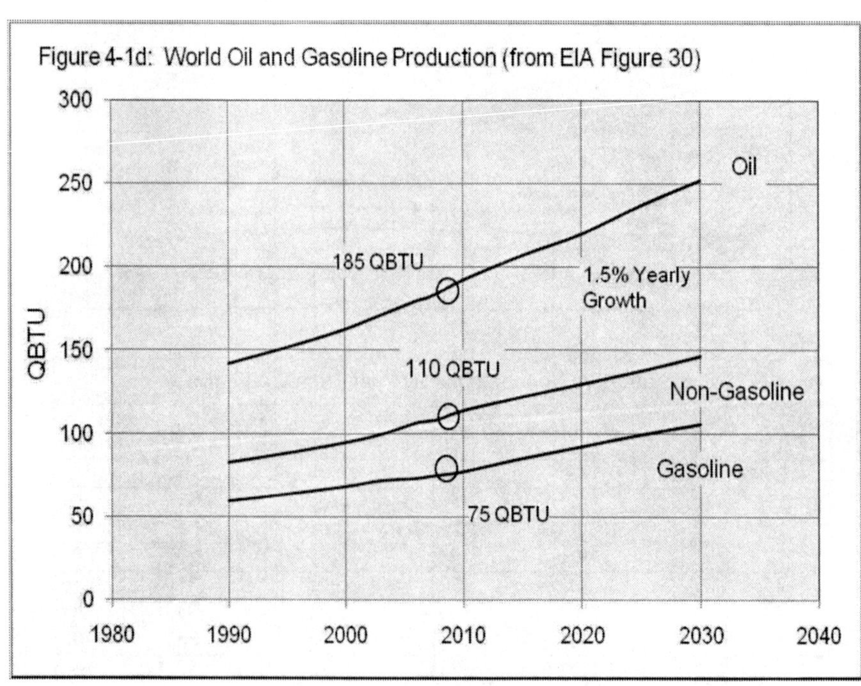

## Section 11: Reference Data and Figures

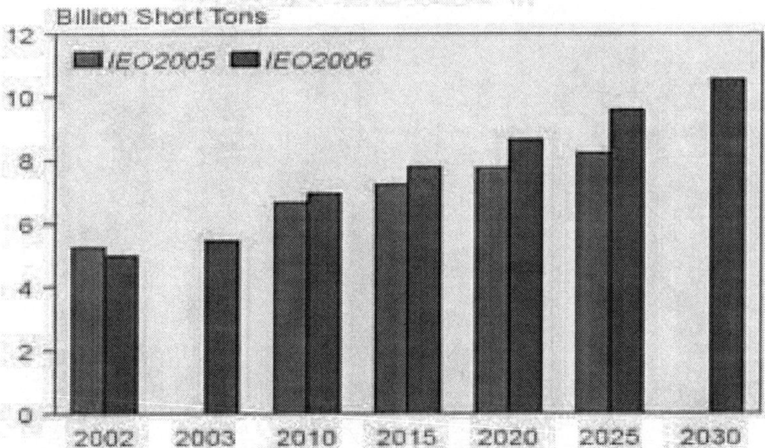


Author's Graphical Data Reduction:

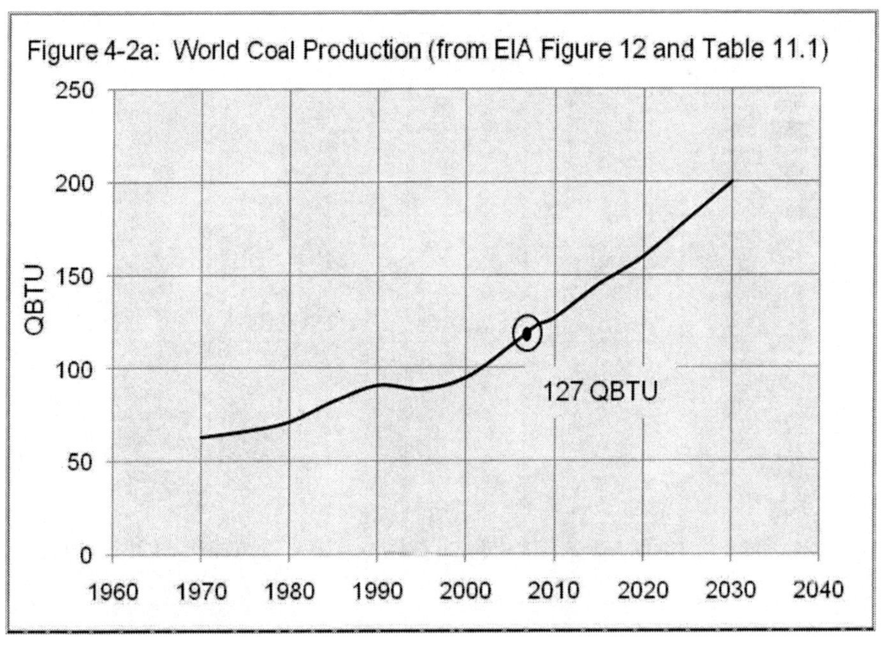

## Section 11: Reference Data and Figures

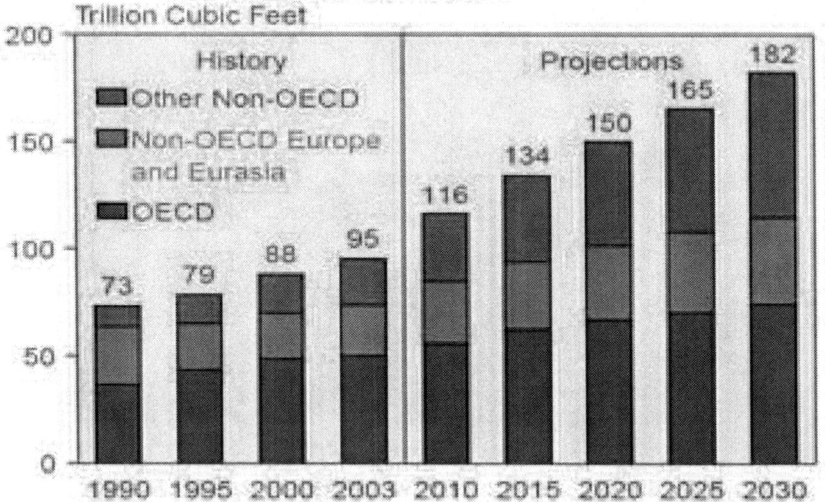


Author's Graphical Data Reduction:

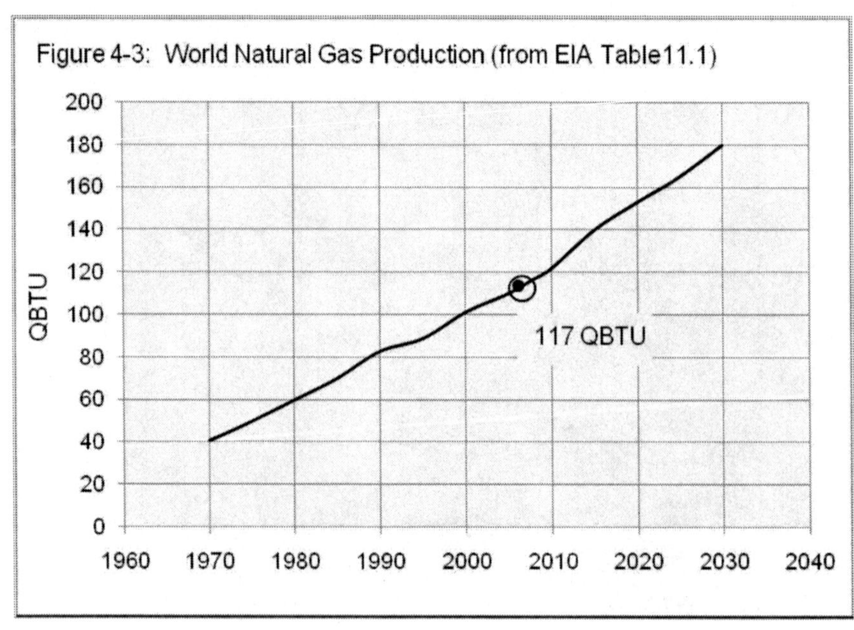

## Section 11: Reference Data and Figures

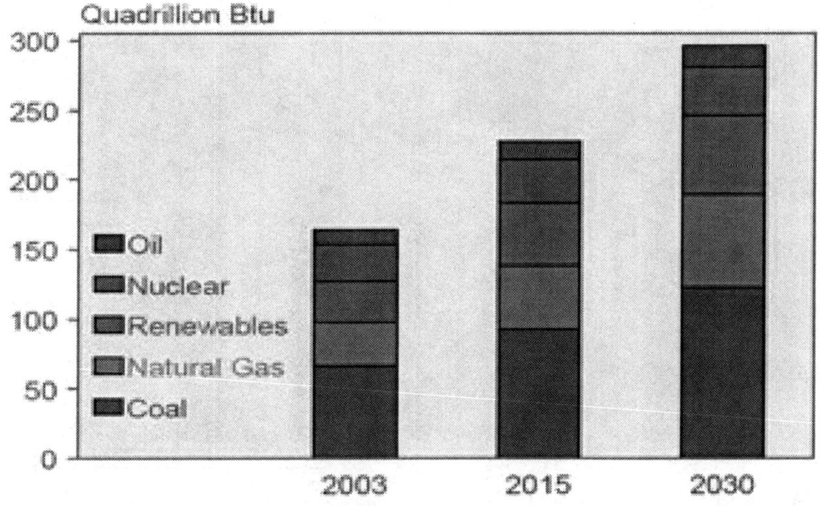


Author's Graphical Data Reduction:

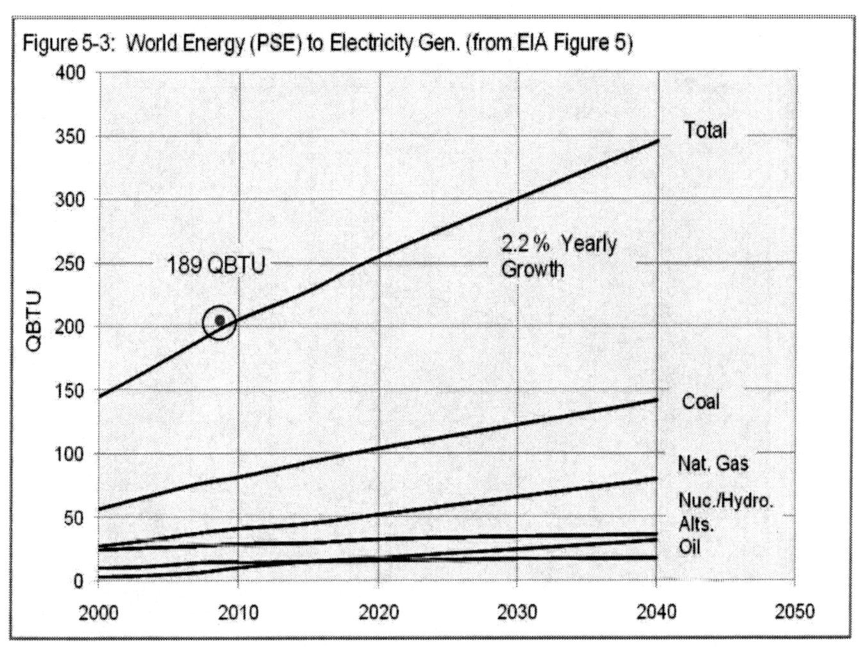

## Section 11: Reference Data and Figures

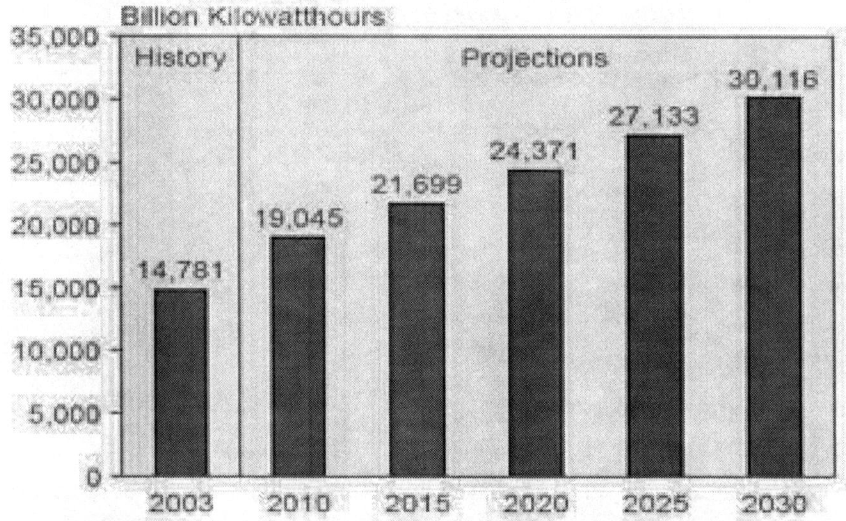


Author's Graphical Data Reduction:

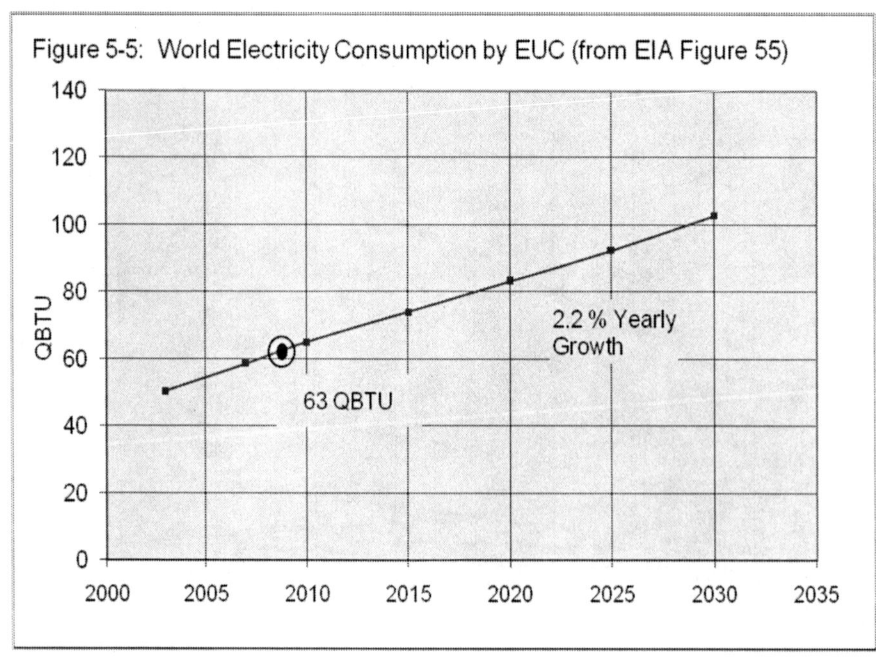

## Section 11: Reference Data and Figures

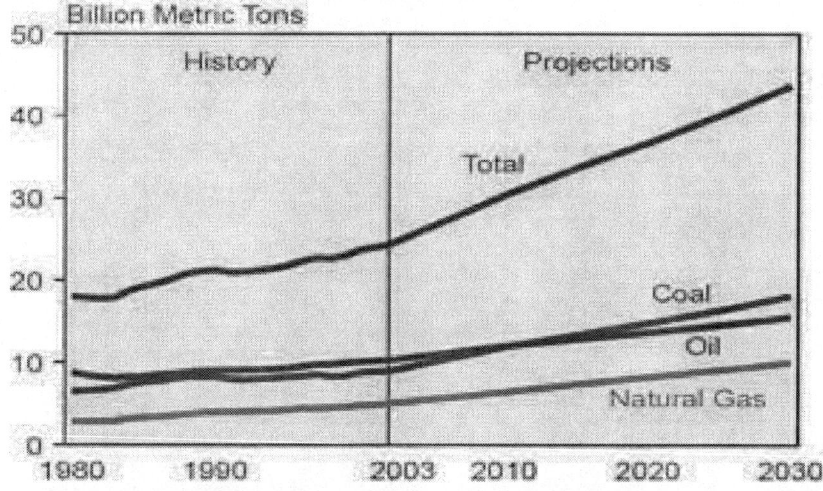


Author's Graphical Data Reduction:

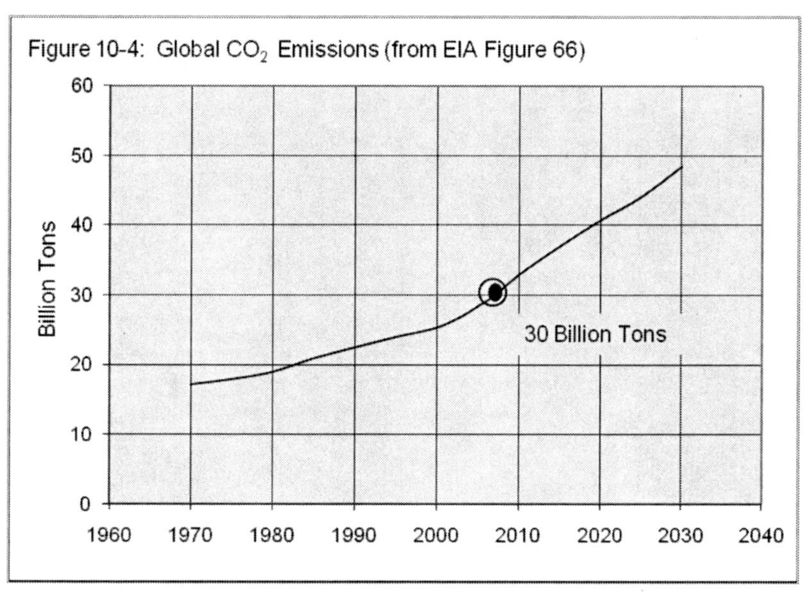

LaVergne, TN USA
30 December 2009

168495LV00003B/3/P

9 781935 125105